Decentralization in Infinite Horizon Economies

Decentralization in Infinite Horizon Economies

EDITED BY

Mukul Majumdar

Routledge
Taylor & Francis Group
LONDON AND NEW YORK

First publishing 1992 by Westview Press, Inc.

Published 2018 by Routledge
52 Vanderbilt Avenue, New York, NY 10017
2 Park Square, Milton Park, Abingdon, Oxon OX14 4RN

Routledge is an imprint of the Taylor & Francis Group, an informa business

Library of Congress Cataloging-in-Publication Data
Decentralization in infinite horizon economies / edited by Mukul Majumdar.
p. cm.
ISBN 0-8133-8090-1
1. Decentralization in management. 2. Decision-making. 3. Resource allocation. 4. Mathematical optimization. I. Majumdar, Mukul, 1944–
HD50.D44 1992
339.4'2—dc20 **91-45889**
CIP

ISBN 13: 978-0-367-01594-7 (hbk)

Contents

Foreword

How can Pareto efficiency be achieved by a decentralized price system which operates through time with no definite terminal date? After forty years of research, the problems raised by this question are not exhausted. This book presents a set of recent articles dealing with the issue and very usefully contributes to our understanding of it, and I am glad to greet its publication.

The challenging difficulties motivating these articles appear when the full extent of decentralization of economic activities is recognized in an infinite horizon framework. Since the early times when the problem was first considered in mathematical economics, skepticism about its significance was expressed: Could the shift from a finite to an infinite horizon have serious implications for the mathematical properties of resource allocation theory or for their interpretation? Intuition gave a negative answer, so that many people too easily claimed that the problems had been somehow solved; a loose reference was then given to one paper or the other that really dealt only with a related question but not with the hard ones. This still happens occasionally.

It would, of course, be still less justified to claim that the virtues of the price system stand or fall depending on whether one looks at a finite or an infinite horizon. We must, however, agree with a more modest but stubborn proposition, namely that we are facing one of those cases in which scientists have to feel uncomfortable because the validity of some central propositions in their discipline remains exposed to doubt with respect to some not fully negligible circumstances. This lack of comfort is the true reason for digging deeper into the pending issues, as is done by the various chapters in this book.

Mukul Majumdar provides in the introductory chapter an excellent presentation of the issues raised by the contributors and their results. He is obviously much more qualified for such a task than I, as I have not worked on the subject for so many years. I may, however, be permitted to signal to the reader one development that, among others, looks to me as particularly worth pursuing. Chapters 8 and 9 explore the concept of an evolutionary process, in which agents make rolling plans up to a finitely distant horizon and periodically revise these plans before reaching their horizon; the constraints at the terminal date of a rolling plan are then somehow related to information available at its initial date. This framework appears the most relevant one to study. Since these two chapters, respectively, give a negative result and a positive one, there is ample room for further research, aiming in particular at generalizing the positive result and at closing the gap with the negative one.

Edmond Malinvaud
College de France, Paris

Acknowledgments

In 1968 I was a research assistant in a project directed by Daniel McFadden and Roy Radner at the University of California–Berkeley. Both were generous in sharing time with me to discuss outstanding theoretical issues in intertemporal economics as was Avinash Dixit. My attention was drawn to the classic paper of Edmond Malinvaud and the role of his "transversality condition" in signalling inefficiency due to capital overaccumulation in a decentralized competitive economy. In 1975, when John Chipman invited me to give a seminar at the University of Minnesota, Leonid Hurwicz speculated on a more general question on the possibility of attaining intertemporal efficiency or optimality by designing some decentralized—but not necessarily competitive—resource allocation mechanism. Collaboration with Leonid Hurwicz started in 1977 and eventually led to the essays collected in this volume. It has been a pleasure to interact with the contributors.

Over the last twenty years, I have had the good fortune of collaborating with Tapan Mitra. His insights into the structure of dynamic models have been exceptionally helpful in deriving definitive results that I could not have proved on my own. In this project, too, he has played an exceptional role.

Edmond Malinvaud's influence on my research defies adequate acknowledgment. I have also benefitted from the important studies on growth and capital theory by David Cass, David Gale, Bezalel Peleg, and David Starrett.

Some of the essays in this collection appeared in a symposium in *The Journal of Economic Theory* (volume 45, number 2, August 1988) that I organized. In that connection I am most grateful to Karl Shell for his help and encouragement. For giving me permission to reprint, I would like to thank Academic Press and Springer Verlag.

Financial support from the National Science Foundation, the Guggenheim Foundation, and Cornell University has enabled me to take a long-run view of research. Indeed, successive chairs of the Economics Department—T.C. Liu (who took a chance on me), Erik Thorbecke, Kenneth Burdett, and David Easley—contributed to the creation of an environment that enabled me to pursue a single theme over a long period.

For preparation of this volume I owe much to Ann Stiles and to my research assistant, Manjira Roychoudhury. The project would not have materialized without the enthusiasm of Spencer Carr.

Mukul Majumdar

1

Decentralization in Infinite Horizon Economies: An Introduction*

MUKUL MAJUMDAR

Department of Economics, Cornell University, Ithaca, New York 14853

The paper summarizes some issues involved in developing a theory of decentralized resource allocation mechanism in infinite horizon economies. Even in a classical environment, a competitive mechanism may lead to an inefficient or non-optimal allocation for an infinite horizon economy. It remains, however, to explore whether the usual price-quantity messages can be supplemented to design decentralized mechanisms generating optimal outcomes. *Journal of Economic Literature* Classification Numbers: 022, 027, 111, 113.

I. INTRODUCTION

The collection of papers in this Symposium represents a confluence of two streams of ideas. The first stream flows from Malinvaud's analysis [20] of efficient accumulation (and "efficiency pricing") in infinite horizon economies and the subsequent development of the theory of optimal growth (and "competitive pricing") by David Gale [6, 7] and others. The second stream flows from the formalization of "informational decentralization" (and related isssues) beginning with Hurwicz [10]. In what follows I shall outline some questions, provide some partial answers, and indicate a few directions that need further exploration in the theory of decentalization for dynamic economies with unbounded time horizons.

* This is a revised version of the working paper (No. 8705) circulated by the Center for Analytic Economics at Cornell. Research support from the Center, the National Science Foundation, and the Warshow endowment at Cornell is gratefully acknowledged. For detailed suggestions on revisions, I am grateful to Karl Shell. I thank W. Brock, L. Hurwicz, T. Mitra, D. Easley, M. Kurz, D. Ray, and E. Malinvaud for comments at various stages. Roy Radner introduced me to infinite horizon efficiency pricing problems and the transversality condition. This essay is dedicated to the memory of Nirmal Kanti Majumdar.

II. The Framework

Failure of the fundamental theorems of welfare economics in infinite horizon economies (that satisfy the "usual" assumptions of convexity and lack of externalities) has been the subject of extended discussion.[1] Malinvaud's paper [20] exposed two startling possibilities. First, as intertemporally efficient infinite program need not have a supporting system of competitive prives [he introduced the additional "non-tightness" condition to establish the existence of such prices]. Second, intertemporal profit maximization relative to a system of strictly positive prices need not guarantee efficiency [he introduced a "transversality condition" in the form (2.3b) stated below to derive a sufficient condition that rules out capital overaccumulation which is at the root of inefficiency of competitive programs]. Going beyond intertemporal efficiency, the fact that a competitive equilibrium allocation is not necessarily Pareto optimal was observed by Samuelson in the framework of an infinite horizon model with overlapping generations.[2] There are an infinite number of (dated) commodities in these models (and an infinity of agents in the model with overlapping generations). One should recall the results on characterizing efficiency in terms of present value maximization (see Radner [25]) and on the optimality of a valuation equilibrium (see Debreu [4]) in models where a price system is defined as a linear functional on an infinite-dimensional commodity space. It was recognized that such an approach led to difficulties in developing a theory of price-guided allocations in a *decentralized* economy [since a linear functional need not generate a price system for every period].[3]

[1] Without doing justice to the contributors, I can refer to the excellent expositions in Koopmans [14], Starrett [27], and Shell [26] on basic issues of intertemporal allocation, with an infinite horizon. On decentralization, the expositions in Koopmans [14] and Hurwicz [9, 11] provide the background material.

[2] The framework is also in Malinvaud [20], but did not play a major role. Of interest is the recent acknowledgment to the work of Allais by Malinvaud [21]. Samuelson's stimulating paper served as the point of departure for a voluminous literature dealing with a variety of issues in monetary and capital theory. Of particular importance are the paper by Gale [8] and the collection adited by Kareken and Wallace [13]. The scope of this essay (and the symposium) is restricted to decentralization in non-monetary models.

[3] It was noted by Radner [25] that even a continuous linear functional on an infinite-dimensional commodity space need not provide a system of prices for each period. Although the "value" of a consumption program can be given some interpretation, it is not clear how such continuous linear functionals can be used in achieving *decentralization* of decisions. Relationship between the approaches by Malinvaud and Debreu was studied in Radner [25] and Majumdar [16]. Explicit recognition of the difficulties involved in interpreting linear functionals in a *decentralized* economy can be found in a large number of papers on economies with an infinite-dimensional commodity space (e.g., Shell [26] asked "what is the economic interpretation to be given to a separating hyperplane that does not provide a price vector for the decentralized economy?").

Indeed, decentralization was a prominent theme in the classic essay ("Allocation of Resources and the Price System") of Koopmans [14]. He first employed two simple examples to illustrate the possibilities of using prices to decentralize economic decisions and then reviewed some significant developments in both static and dynamic theories of resource allocation with a recurrent emphasis on the role of competitive prices in achieving "informational simplification" and independent decision making. His first example was the Robinson Crusoe economy, in which he showed how production and consumption decisions could be separated. In this second example, Koopmans showed how aggregate profit maximization could be achieved by a group of independent enterpreneurs maximizing individual profits with prices treated as parameters. Both these examples are relevant for interpreting the formal results reported in the papers collected here. After some formal definitions are introduced, I shall recall the insightful remarks in Koopmans [14] on the basic problem of attaining efficient or optimal allocations in *infinite horizon economies* through a decentralized system of decision making. Interestingly enough, the resolution of this problem has remained an open question since then.

II.a. *Stationary Environment*

Consider an infinite horizon, one-good model with a time-invariant *technology* described by a twice differentiable gross output function f which has the standard monotonicity and strict concavity properties and in addition, satisfies Inada conditions.[4] Given the initial stock $y_0 > 0$, a feasible program of resource allocation [briefly, "a feasible $(\mathbf{x}, \mathbf{y}, \mathbf{c})$"] from y_0 consists of non-negative sequences of inputs $\mathbf{x} = (x_t)$, stocks $\mathbf{y} = (y_t)$, and consumptions $\mathbf{c} = (c_t)$ satisfying

$$\begin{aligned} &x_0 + c_0 = y_0 \\ &x_t + c_t = y_t \qquad \text{for all} \quad t \geqslant 1. \\ &y_t = f(x_{t-1}) \end{aligned} \tag{2.1}$$

Alternative feasible programs are evaluated according to some welfare criterion. Let δ be the discount factor satisfying $0 < \delta \leqslant 1$, and u be the one-priod *felicity* functions assumed to be twice differentiable, strictly

[4] To be precise, $f: R_+ \to R_+$ is continuous on R_+; $f(0) = 0$; $f(x)$ is twice differentiable at $x > 0$, with $f'(x) > 0$, $f''(x) < 0$; moreover, (a) $\lim_{x \downarrow 0} f'(x) = \infty$, and (b) $\lim_{x \uparrow \infty} f'(x) = 1 - \theta_2$, $0 < \theta_2 < 1$. In what follows, this collection of conditions is referred to as "conditions (F)."

increasing, strictly concave and to satisfy an Inada condition.[5] When $\delta < 1$ (resp. $\delta = 1$) we have the *discounted* (resp. *undiscounted*) case. A feasible $(\mathbf{x}^*, \mathbf{y}^*, \mathbf{c}^*)$ from y_0 is optimal if

$$\limsup_{T \to \infty} \sum_{t=0}^{T} \delta^t [u(c_t) - u(c_t^*)] \leqslant 0 \tag{2.2}$$

for all feasible $(\mathbf{x}, \mathbf{y}, \mathbf{c})$ from y_0.

As in static models, the role of prices in guiding optimal resource allocation has been studied formally through the notion of a competitive program. A feasible $(\mathbf{x}, \mathbf{y}, \mathbf{c})$ is *competitive* if there exists a nonzero sequence $\mathbf{p} = (p_t)$ of (discounted) prices such that for all $t \geqslant 0$

$$\delta^t u(c_t) - p_t c_t \geqslant \delta^t u(c) - p_t c, \qquad \text{for all} \quad c \geqslant 0 \tag{G}$$

$$p_{t+1} f(x_t) - p_t x_t \geqslant p_{t+1} f(x) - p_t x, \qquad \text{for all} \quad x \geqslant 0. \tag{M}$$

The condition (G) is that of expenditure minimization and (M) is the condition of intertemporal profit maximization relative to the supporting price system $\mathbf{p} = (p_t)$.[6] Given the assumptions on the technology and the felicity functions [conditions (F) and (U) in footnotes 4 and 5], the supporting price system $\mathbf{p}$ and the quantity sequences $\mathbf{x}$, $\mathbf{y}$, $\mathbf{c}$ of a competitive program are all strictly positive. Our interest in competitive programs is naturally due to the following:

PROPOSITION 1. *A feasible* $(\mathbf{x}, \mathbf{y}, \mathbf{c})$ *from* $y > 0$ *is optimal if and only if it is competitive relative to* $\mathbf{p} = (p_t)$ *and*

$$(p_t x_t) \text{ is bounded} \qquad [\text{when } \delta = 1] \tag{2.3a}$$

$$\lim_{t \uparrow \infty} p_t x_t = 0 \qquad [\text{when } \delta < 1]. \tag{2.3b}$$

The rough equivalence summarized in the proposition reminds us of the two fundamental theorems of "new" welfare economics (or the twin properties of non-wastefulness and unbiasedness of the competitive mechanism in the language of Hurwicz [11]) and—as in the static model—one turns to the use of prices in decentralization and in achieving informational

[5] Formally, u is defined either on R_+ or R_{++} with values in R. When u is defined only on R_{++}, we assume that $\lim_{c \downarrow 0} u(c) = -\infty$; $u(c)$ is twice differentiable at $c > 0$ with $u'(c) > 0$, $u''(c) < 0$ and $\lim_{c \downarrow 0} u'(c) = \infty$, $\lim_{c \uparrow \infty} u'(c) = 0$. This collection of conditions is subsequently referred to as "conditions (U)."

[6] Derivation of prices [satisfying (M)] associated with *efficient* programs has been the contribution of Malinvaud [20] and in a number of papers beginning with Gale [6, 7] competitive prices supporting *optimal* programs were derived.

efficiency. Paraphrasing from Koopmans [14], one can suggest that "by giving sufficiently free rein to our imagination, we can visualize" the Gale–Malinvaud conditions (G) and (M) "as being satisfied through a decentralization of decisions" among an infinite number of agents, each in charge of a part of the program in a particular period. One can directly adapt the *verification scenario* of Hurwicz [9] to emphasize separation of consumption and production decisions period after period: in period t, the expenditure-minimization condition (G) is verified given (c_t, p_t) *only* on the basis of the preferences represented by $(\delta^t u)$ and the intertemporal profit maximization condition (M) is verified given $(p_t, p_{t+1}, x_t, y_{t+1})$ only on the basis of the technology (the function f). The verification involves a finite number of messages, and preserves "privacy." However, to my knowledge it was Koopmans [14] who first noted that "even at this level of abstraction it is difficult to see how the task of meeting" condition (2.3a) or (2.3b) "could be pinned down to any particular decision maker. This in a new condition to which there is no counterpart" in finite horizon models. Note that if the producer in period t is allowed to observe a *finite* number of prices and quantities it cannot verify (2.3a) or (2.3b).[7]

It is fair to say that the point of departure of this symposium is really an attempt *to design "decentralized" resource allocation mechanisms that, unlike the competitive mechanism, can detect long-run inefficiency or non-optimality on the basis of a sequence of "verifications."*

II.b. *Possibility of Decentralization*

One of the basic ideas pursued in the papers by Brock and Majumdar [1] and Dasgupta and Mitra [2, 3] is to *supplement* the conditions (G) and (M) by a period-by-period verification rule that replaces (2.3a) or (2.3b). As an illustration of this approach consider the unique positive solution x_δ^* to the equation

$$\delta f'(x) = 1 \tag{2.4}$$

and let $y_\delta^* = f(x_\delta^*)$, $c_\delta^* = y_\delta^* - x_\delta^*$, and $p_t^* = \delta^t u'(c_\delta^*)$. The triplet $(x_\delta^*, y_\delta^*, c_\delta^*)$ represents the golden rule input, stock, and consumption. It is well known that the stationary program defined by $x_t^* = x_\delta^*$, $y_t^* = y_\delta^*$, $c_t^* = c_\delta^*$ is the unique optimal program from $y_\delta^* > 0$ and that it is competitive relative to prices $\mathbf{p}^* = (p_t^*)$.

Now suppose that $(\bar{\mathbf{x}}, \bar{\mathbf{y}}, \bar{\mathbf{c}})$ is the unique optimal program from some

[7] Commenting on Malinvaud's transversality condition [see (2.3b)] to guarantee production efficiency, Kurz and Starrett [15] also emphasized that such conditions are not "decentralizable for the myopically behaving firm. Thus, our problem could be reduced to the fact that we do not really have a decentralizable rule that will ensure the efficiency of profit-maximizing allocation."

$\bar{y}_0 > 0$. Recall from Mitra [22] that $\bar{y}_t$ ($\gtreqless$) y^*_δ and $\bar{c}_t$ ($\gtreqless$) c^*_δ for all $t \geqslant 0$ if and only if $\bar{y}_0 \gtreqless y^*_\delta$. Suppose $\bar{\mathbf{p}} = (\bar{p}_t)$ is the price system relative to which $(\bar{\mathbf{x}}, \bar{\mathbf{y}}, \bar{\mathbf{c}})$ satisfies (G) and (M). Then (since $\bar{c}_t > 0$ for all t) $\bar{p}_t = \delta^t u'(\bar{c}_t)$ and one can show that

$$(\bar{y}_t - y^*_\delta)(\bar{p}_t - p^*_t) \leqslant 0 \qquad \text{for all} \quad t \geqslant 0. \tag{2.5}$$

Now observe that the inequality corresponding to (2.5) can be verified by the producer in period $t-1$. The important feature is that if for any competitive program $(\bar{\mathbf{x}}, \bar{\mathbf{y}}, \bar{\mathbf{c}})$ one obtains (2.5) for all t, one can conclude that it is optimal. Formally, we can summarize this as:

PROPOSITION 2. *A feasible* $(\bar{\mathbf{x}}, \bar{\mathbf{y}}, \bar{\mathbf{c}})$ *from* $\bar{y}_0 > 0$ *is optimal if and only if it is competitive relative to* $\bar{\mathbf{p}} = (\bar{\mathbf{p}}_t)$ *and*

$$(\bar{\mathbf{y}}_t - y^*_\delta)(\bar{p}_t - p^*_t) \leqslant 0 \qquad \textit{for all} \quad t \geqslant 0. \tag{S}$$

Strictly speaking, the above proposition is not covered by any of the formal results reported in this symposium [given the conditions (F) and (U) in footnotes (2) and (3)].[8]

The "possibility result" suggested in Proposition 2 immediately leads to the question of its robustness. First, suppose that we confine ourselves to the one-good structure. Then other "period-by-period" characterizations of optimality are certainly possible. For example, one can establish:

PROPOSITION 3. *A feasible* $(\mathbf{x}, \mathbf{y}, \mathbf{c})$ *from* $y_0 > 0$ *is optimal if and only if it is competitive and*

$$x_{t+1} (\gtreqless) x_t \quad \textit{whenever} \quad x_t (\lesseqgtr) x^*_\delta \quad \textit{for all} \quad t \geqslant 0. \tag{2.6}$$

Instead of the additional verification suggested by condition (S), the producer is asked to check the additional condition (2.6). Again, this can be done in period t with the knowledge of a *finite number of prices and quantities* [although a producer in period t is now required to know the quantities in two adjacent periods t, $t+1$]. It is of interest to note that even when the production function f has a phase of increasing returns (i.e., we are in a "non-classical" encironment), optimal programs can be shown to enjoy properties like (2.6) (see, e.g., the paper by Majumdar and Nermuth [18]). This holds out the possibility of intertemporal decentralization in such models.[9]

We now turn to multisector models. Unfortunately, the monotonicity result (2.6) is a feature that only generally holds in the one-good framework, and there is no immediate clue to the problem of extending

[8] A sketch of an elementary proof [which deals with both the discounted and undiscounted cases in a single framework] of Propositions 2 and 3 is available in the working paper (No. 87-05) circulated by the Center for Analytic Economics at Cornell.

[9] This is one of the themes of Majumdar and Ray [19].

Proposition 3 in a multisectoral framework. However, it turns out that Proposition 2 can indeed be extended to a class of multisector models. Such an extension to an "open model" is the principal theme of three of the accompanying papers [1, 2, 23]. The golden rule program plays a critical role in formulating the condition (S).

In a *closed model* (i.e., one without primary factors), such an optimal *stationary* program does not exist. However, an activity analysis model with no joint production (the "simple linear production model" of David Gale [5, p. 294]) is examined in Dasgupta and Mitra [3]. They have made considerable progress by using an efficient balanced growth program as a possible benchmark [although the precise result is more limited in scope than the multisectoral version of Proposition 2 in open models]. Absence of joint production means that the model is really capturing circulating rather than durable capital, and it is of interest to seek further generalizations to von Neumann-type models.

We may conclude, then, that in a wide class of stationary, infinite horizon economies, the condition (S) makes it possible to apply Hurwicz's verification scenario. A second question raised by the possibility results is related to incentives. In the context of static welfare economics it is argued that behavior rules like utility maximization or profit maximization "utilize incentives" that are naturally operative in the market. Leaving aside the point that Robinson Crusoe-type models [which are used for expository ease] are inappropriate for analyzing maximizing behavior accepting prices, and perhaps the more important point that profit or utility maximization need not completely or accurately describe the incentives governing human action, it is certainly true that irrespective of the number of agents, conditions like (2.6) or the (S) condition do not have any immediate link to incentives, and this surely is an important area of future research.

Finally, recall that Malinvaud's original contribution incorporated multiple firms/producers in each period, and [treating the aggregate production possibility set as a sum of independent individual production possibilities] established an *equivalence* between aggregate profit maximization and individual profit maximization that had a clear interpretation in terms of decentralization within a particular period (this, of course, is related to the second example in Koopmans, as mentioned above). The papers in the symposium sidestep this (basically atemporal) multifirm issue, but it is not clear how such an equivalence can be established between the aggregate level and the individual level with respect to the condition (S)[10] [or (2.6)].

[10] If there are *m* identical firms in any given period, sufficiency of (S) can still be proved, as pointed out by Modecai Kurz.

II.c. *Uncertainty*

A further extension of the (S) characterization of optimality from a deterministic to a stochastic framework (in the *undiscounted* case) is achieved in Nyarko [23]. The particular setup of production that he has chosen is by now one of a few rigorous approaches to modelling decision making under uncertainty. Given the unavoidable technical complexities in a formal treatment, one may lose sight of the primary motivation behind introduction of undertainty in the present context. Here, again, I would like to recall the interesting conjecture of Koopmans: "It may be well to remind ourselves of the artificiality of studying one aspect of time—its indefinite duration—without another equally essential aspect—incomplete information about the future. It may be conjectured, for instance, that the introduction into the model of a combination of uncertainty about the future technology and risk aversion on the part of the producers would promptly close off the loopholes" [created by (2.3a) and (2.3(b)] and guarantee long-run optimality. I should stress that subsequent research by Zilcha [28, 29] provided us with a complete characterization analogous to Proposition 1 which could not "circumvent" the appropriate stochastic version of a transversality condition [which Nyarko does]. However, no attempt has been made so far to identify formal conditions under which the Koopmans conjecture will be valid. Moreover a routine interpretation of the duality results on optimality and supporting prices in terms of state-dependent commodities should remind one of the complaints of Radner [24][11] that serious rethinking is needed to provide a coherent and tractable account of decentralization in a dynamic, infinite horizon world subjected to random shocks.

III. Non-Stationary Environments

A general framework for a formal treatment of issues related to decentralization in a dynamic world is suggested in Hurwicz and Majumdar [12]. One has to specify what the designer of the mechanisms knows[12]

[11] "If the Arrow–Debreu model is given a literal interpretation, then it clearly requires that the economic agents possess capabilities of imagination and calculation that exceed reality by many orders of magnitude. Related to this is the observation that the theory requires in principle a complete system of insurance and futures markets, which appears too complex, detailed and refined to have practical significance."

[12] In the stationary case discussed above, the designer knows that the sequences of production ($\mathbf{f}$) and utility ($\mathbf{u}$) functions are stationary, i.e., $\mathbf{f} = (f^{(\infty)})$, $\mathbf{u} = (u^{(\infty)})$, where f and u satisfy (F) and (U), respectively. It does *not* know the particular f or u nor does it know the initial stock of the discount factor.

and what the agents in each period are allowed to know.[13] Two observations are needed to contrast the approach in [12] with the literature on decentralization for the static case. Most of the analytical results are derived with the assumption that the verification rule in period t is based on a partial history of the environment; in particular, we allow agents in period t to use the knowledge of the environment from "the beginning of history" up to a finite number of periods into the future. Thus, the focus is on *limited* information rather than a strict "preservation of privacy." Second, the class of mechanisms analysed is restricted to those of a "proposed program" type, i.e., that do not *explicitly* use auxiliary variables such as prices. However, as agents are allowed to respond on the basis of past *and* current technologies, this class is rich enough to include equivalents of competitive mechanisms (given the special structures of the one-good models).

Two evaluation criteria are formally examined in some detail: (Malinvaud's) intertemporal efficiency (see Section III.1 of Hurwicz and Majumdar [12]) and the discounted case with a time-invariant felicity function (recall (2.6) with $\delta < 1$). With regard to the first criterion, an impossibility result is obtained: it is impossible to design a sequence of temporally decentralized verification rules that is both nonwasteful and unbiased [achieving only one of these properties is possible]. A second set of impossibility results is presented for the discounted optimality case. If the set of admissible technologies consists exclusively of *linear* technologies (and no further restriction on variation over this set is imposed), it is impossible to design a sequence of temporally decentralized verification rules that guarantees optimality excepting the case when the felicity function is logarithmic (see (T.5) in [12]).

If the class of admissible technologies is allowed to include a pair of strictly concave production functions with a specific property (namely, that at some positive input level both yield the *same* output but have different marginal productivities[14]), it is, again, impossible to design a sequence of temporally decentralized verification rules (T.5.2) in [12]). The key element is *incomplete* information about the possibility of future technological change.

I should add that an impossibility result analogous to (T.5.2) in [12]

[13] In the stationary case, the "consumer" in period t knows $(\delta^t u)$, and the messages in terms of prices and quantities that the mechanism allows it to observe; the "producer" in period t knows f and the relevant price-quantity messages. In the mechanism I described earlier, their role is to verify some inequalities.

[14] Note the following implication regarding (A.3.5) in Hurwicz and Majumdar [12]: if the set of admissible gross output functions does not satisfy (A.3.5) it is "thin" (has empty interior relative to a widely used topology).

continues to hold (under a similar set of assumptions) even when one uses the "undiscounted" overtaking criterion (recall (2.6) *with* $\delta = 1$)). The relevant details are spelled out in Majumdar [17].

IV. Concluding Comments

Interesting links between Pareto optimality and intertemporal efficiency have been established in various models, and the research reported in this Symposium can be viewed as a first step towards exploring the question of decentralization in other models of intertemporal equilibrium. Given the impossibility results involving competitive models (with a tatonnement scenario in the wings), it may well be that one can best hope for some "second best" properties of informationally decentralized systems in which the agents recognize their finite life spans and limited computational abilities and attempt to formulate and execute plans without waiting for an eventual equilibrium. Developing such genuinely dynamic mechanisms is perhaps the most challenging direction for future research.

References

1. W. A. Brock and M. Majumdar, On characterizing optimal competitive programs in terms of decentralizable conditions, *J. Econ. Theory* **45** (1988), 262–273.
2. S. Dasgupta and T. Mitra, Characterization of intertemporal optimality in terms of decentralizable conditions: The discounted case, *J. Econ. Theory* **45** (1988), 274–287.
3. S. Dasgupta and T. Mitra, Intertemporal optimality in a closed linear model of production, *J. Econ. Theory* **45** (1988), 288–315.
4. G. Debreu, Valuation equilibrium and Pareto optimum, *Proc. Nat. Acad. Sci. U.S.A.* **40** (1959), 588–592.
5. D. Gale, "The theory of Linear Economic Models," McGraw–Hill, New York, 1960.
6. D. Gale, On optimal development in a multi-sector economy, *Rev. Econ. Stud.* **34** (1967), 1–18.
7. D. Gale, A geometric duality theorem with economic applications, *Rev. Econ. Stud.* **34** (1967), 19–26.
8. D. Gale, Pure exchange equilibrium of dynamic economic models, *J. Econ. Theory* **6** (1973), 12–36.
9. L. Hurwicz, On informational decentralization and efficiency in resource allocation mechanisms, *in* "Stud. Math. Econ., MAA Stud. Math." (S. Reiter, Ed.), Vol. 25, pp. 238–350, Math. Assoc. Amer., Washington, D.C., 1986.
10. L. Hurwicz, Optimality and informational efficiency in resource allocation processes, *in* "Mathematical Methods in the Social Sciences" (Arrow, Karlin, and Suppes, Eds.), Stanford Univ. Press, Stanford, CA, 1960, 27–46.
11. L. Hurwicz, The design of resource allocation mechanisms, *Amer. Econ. Rev.: Papers and Proceedings*, **58** (1973), 1–30.
12. L. Hurwicz and M. Majumdar, Optimal intertemporal allocation mechanisms and decentralization of decisions, *J. Econ. Theory* **45** (1988), 228–261.

13. J. H. Kareken and N. Wallace (Eds.), "Models of Monetary Economics," Federal Reserve Bank of Minneapolis, 1980.
14. T. C. Koopmans, "Three Essays on the State of Economic Science," McGraw–Hill, New York, 1957.
15. M. Kurz and D. Starrett, On the efficiency of competitive program in an infinite horizon model, *Rev. Econ. Stud.* **38** (1970), 571–589.
16. M. Majumdar, Some approximation theorems on efficiency prices for infinite programs, *J. Econ. Theory* **2** (1970), 399–410.
17. M. Majumdar, "Optimal Intertemporal Allocation Mechanisms and Decentralization: The Undiscounted Case," Working No. 393, Department of Economics, Cornell University, 1987.
18. M. Majumdar and M. Nermuth, Dynamic optimization in non-convex models with irreverisible investment: Monotonicity and turnpike results, *Z. National ö konom.* **42** (1982), 339–362.
19. M. Majumdar and D. Ray, Some aspects of intertemporal decentralization and non-convexity, forthcoming.
20. E. Malinvaud, Capital accumulation and efficient allocation of resources, *Econometrica* **21** (1953), 233–268.
21. E. Malinvaud, The overlapping generations models in 1947, *J. Econ. Lit.* **25** (1987), 103–105.
22. T. Mitra, On optimal economic growth with variable discount rates, *Int. Econ. Rev.* **20** (1979), 133–145.
23. Y. Nyarko, On characterizing optimality of stochastic competitive processes, *J. Econ. Theory* **45** (1988), 316–329.
24. R. Radner, Problems in the theory of markets under uncertainty, *Amer. Econ. Rev. Papers and Proceedings*, (1970), 454–460.
25. R. Radner, Efficiency prices for infinite horizon production programmes, *Rev. Econ. Stud.* **34** (1967), 1–66.
26. K. Shell, Notes on the economics of infinity, *J. Polit. Econ.* **79** (1971), 1002–1011.
27. D. Starrett, "Contributions to the Theory of Capital in Infinite Horizon Models," Technical Report No. 16, IMSSS, Stanford University, 1968.
28. I. Zilcha, Characterization by prices of optimal programs under uncertainty, *J. Math. Econ.* **3** (1976), 173–183.
29. I. Zilcha, Transversality condition in a multi-sector economy under uncertainty, *Econometrica* **44** (1978), 515–525.

2

Optimal Intertemporal Allocation Mechanisms and Decentralization of Decisions*

LEONID HURWICZ

Department of Economics, University of Minnesota, Minneapolis, Minnesota 55455

AND

MUKUL MAJUMDAR

Department of Economics, Cornell University, Ithaca, New York 14853-7601

Problems of achieving an efficient or optimal allocation of resources through a system of decentralized decision-making when the planning horizon is unbounded and the number of agents is infinite have been repeatedly raised, but have not been explored in a formal manner. The objective of this paper is to suggest some points of departure for a systematic investigation of such questions. The main analytical results indicate the impossibility of attaining optimality through intertemporally decentralized mechanisms. *Journal of Economic Literature* Classification Numbers: 022, 024, 111, 113.

I. INTRODUCTION

The question of decentralization of decision-making when the horizon and the number of agents are infinite has been repeatedly raised in the

* The research reported here was started when the authors were visiting the University of California, Berkely, during the academic year 1976–1977. Hurwicz's research was supported by the National Science Foundation Grant SES-8208378, and benefited greatly from the participation in the 1983–1984 program of the Institute for Mathematics and Its Applications at the University of Minnesota, as well as from the appointment at the California Institute of Technology during 1984–1985 as a Sherman Fairchild Distinguished Scholar. Majumdar's research was supported by a fellowship from the John Simon Guggenheim Foundation and by the National Science Foundation Grant SES-8304131. Thanks are due to Professor Tapan Mitra for many insightful conversations and for his help in proving one of the main results. Comments from W. Dean, M. Ali Khan, A. Lewis, J. BenHabib, J. Foster, and F. Hahn were also useful. We are grateful to the referees and editors for their commets and criticisms. Detailed proofs of some results omitted in the interest of brevity are given in Working Papers circulated from Cornell University [5, 6].

literature,[1] but has not so far been explored in a formal, systematic way. It is our objective to suggest some points of departure for such an investigation.

Initially, the question was posed not in the general context of decentralization, but in the more specialized context of optimality and efficiency of *competitive equilibria* over time or of allocation processes based on *intertemporal profit maximization.* In attempts to extend the fundamental theorems of welfare economics, it was noted that the distinction between finite and infinite horizon (with an infinite number of decision-makers) is crucial. In *finite* horizon models, which constitute special cases of a "classical" production environment (with the usual convexity properties), it is well known that every profit maximizing allocation (for definitions, see Koopmans [7, p. 85]) is efficient,[2] and that every efficient allocation can be attained through profit maximization,[3] prices always being treated parametrically. However, as pointed out first by Malinvaud [10], and discussed subsequently by Koopmans [7, p. 111], when the horizon and the number of decision-makers are infinite, a profit maximizing allocation need not be efficient (see Example 3.1 (Pure accumulation programs), and the comments following (3.11) in this paper). To guarantee efficiency, an additional condition—such as "terminal cost minimization" (the condition (b) in Koopmans [7, p. 110] or the "transversality condition" (requiring that the sequence of present values of inputs at profit maximizing prices must converge to zero over time)[4]—becomes indispensable. However, as Koopmans correctly pointed out, while the condition of intertemporal prifit maximization (as usually formulated) is in some sense "decentralized," it is "difficult to see" how the task of meeting conditions like terminal cost minimization or the transversality condition "could be pinned down on any particular decision-maker."

The discussion raises the question whether, in fact, the various "asymptotic" conditions (like the condition (b) of terminal cost

[1] For example, Koopmans [7, p. 111], in discussing condition (b) without which profit maximizing infinite programs may fail to be efficient, says: "... it is difficult to see how the task of meeting condition (b) could be pinned down on any particular decision-maker." This is being contrasted by Koopmans with the fact that the profit maximization condition (a) *can* be "... satisfied through a decentralization of decisions among an infinite number of producers, each in charge of that part of the program relating to one future period." In the literature, condition (b) is called "terminal cost minimization."

[2] For the linear case, see Koopmans [7, p. 87, Prop. 6]. Also, for the general (without-convexity assumptions) non-linear case, see Nikaido [14, p. 185, Theorem 12.1a].

[3] For the linear case, see Koopmans [7, p. 8, Prop. 7]; for the non-linear case, see Nikaido [14, p. 185, Theorem 12.1.b].

[4] For a discussion of the role of the transversality condition in signalling inefficiency due to capital over-accumulation, see Mitra and Majumdar [12, p. 47–48].

minimization in Koopmans [7, p. 110]) are "decentralized," once the latter term has been rigorously defined. If the answer turns out to be negative, the further question is whether or not all efficient and/or optimal allocations can be achieved by *some* informationally decentralized mechanism. Clearly, any resolution of these issues will depend on the concept of decentralization used in the formal analysis. Our definitions were arrived at with the following considerations in mind. First, we are ignoring incentive aspects of decentralization. But it should be remembered that the absence of incentival restrictions on the mechanisms only strenghtens our *impossibility* results. Second, the notion of decentralization should include the competitive mechanism as a special case. Otherwise our results would not cover the possibility of inefficient outcomes pointed out by Malinvaud, and we would be going counter to the tradition veiwing perfect competition as par excellence instance of economic decentralization. Third, we have tried to capture the intuition behind the Koopmans conjecture of the impossibility of decentralization in intertemporal models. This involved classifying as *not* decentralized those mechanisms in which a finite-lived agent can peer arbitrarily far into the future and observe the whole infinite sequence of future messages, but otherwise imposing as few restrictions as possible. In our view the basic feature of the situation is that while the stream of decisions to be made is (denumerably) infinite, as is also the set of decision makers, each (finite-lived!) decision maker makes only a finite number of decisions; he/she has access to only a finite-time interval of observations—typically the present and a subset of the past. Furthermore, as is usual in the study of atemporal models, we assume (the "privacy-preserving" property) that each agent only knows his own characteristic and not those of others. Occasionally we relax even these rather mild conditions and thus obtain stronger *impossibility* theorems. For instance, we permit the agents to have knowledge of not only the past and present but even a (finite or infinite but incomplete) portion of the future.

To analyze the question of (informational) decentralizability in a multiperiod model, whether with finite or infinite horizon, we must consider certain aspects peculiar to such a model.

First, there is the question whether the environment characteristics (technology, preferences) remain constant or change from period to period.

Next, how much information about the environment and its changes over time is assumed available to the designer of the resource allocation mechanism?

The answers depend on the role played by consumer preferences. They do not appear explicitly in the context of efficiency, but do affect solutions to optimization problems in which the designer is assumed to know consumer preferences. We shall see that in such situations environments with stationary technologies do not present interesting problems of decentralizability. For this reason we focus attention on economies with nonstationary technologies.

It may be noted that nontrivial analytical problems do arise even for stationary economies when producers are ignorant of consumer preferences, consumers are ignorant of production technologies, and the designer is ignorant of both—as is natural to assume for a privacy-preserving mechanism. In such situations decentralization may turn out to be impossible with mechanisms of the non-tâtonnement type, but possible with those of the tâtonnement type.[5] By contrast, under conditions of nonstationarity considered in the present paper, and with preferences known or immaterial, decentralization turns out to be impossible even with tâtonnement.

In the discussion that follows we assume that the knowledge of preferences is either unnecessary (as in efficiency) or available to the designer. (Similar implications for decentralizability are obtained by the alternative assumption to the effect that consumer preferences are known to the producer.) Suppose furthermore that the technoogy for period t is expressible as a one parameter family of functions, say $f_t(\cdot\,; \rho_t)$ where f_t is known to the designer but ρ_t may or may not be known, $t=0, 1, \ldots$.

At this point it is useful to see what implicit informational assumptions underlie the Arrow–Debreu model. If one thinks of a distinct firm existing for each $t=0, 1, 2, \ldots$, the Arrow–Debreu model does not require that the system designer have a priori knowledge of the parameters of the various firms. However, it may be taken as known a priori that these parameters vary over specified ranges, say $\rho_t \in \theta_t$. Then the designer's state of information (the a priori admissible class θ) is represented by the Cartesian product $\theta = \theta_1 \times \theta_2 \times \cdots$.

In this context, a meaning attributed to informational decentralization is that the behavioral rule to be followed by firm t does not require that it should know $\rho_{t'}$, for $t' \neq t$. The firm t will, of course, have partial indirect knowledge concerning the $\rho_{t'}$, $t' \neq t$, through the announced prices; however, it need not have any *direct* knowledge of the $\rho_{t'}$.

What is then the appropriate corresponding informational assumption for the designer in a multiperiod model?

First, suppose that the technology is *stationary*, i.e., that $f_t = f_0$ for all t and $\rho_t = \rho_0$ for all t. If this fact is known to the designer, the resulting situation is different from that of the general Arrow–Debreu model: it corresponds to a special case of the latter where it is a priori known to the designer that all firms' technologies are identical, hence that the a priori

[5] See the Introduction to the Symposium for a review of the possibility results in a tâtonnement framework. Hurwicz and Weinberger have recently obtained some impossibility results in a non-tâtonnement context.

admissible class $\theta =$ diagonal of $\theta_1 \times \theta_2 \times \cdots$. In such a situation the privacy-preserving property becomes trivial because knowing one's own ρ_t is equivalent to the knowledge of all other ρ_t since here $\rho_t = \rho_{t'}$ for all t, t'.

The preceding discussion suggests that the analysis of the problem be classified according to some of the possible states of information of the designer, i.e., in our formulation, some of the possible types of a priori admissible sets θ. This may be done as follows:

I. $\theta = \{(\rho_0^*, \rho_1^*, ...)\}$, a singleton for some $\rho_0^*, \rho_1^*, ...$.

I′. Special case: $\rho_t^* = \rho_0^*$ for all t.

II. $\theta = \{(\rho_0, \rho_1, ...): \ \rho_1 = \Psi_1(\rho_0), \ \rho_2 = \Psi_2(\rho_0), ..., \rho_0 \in \theta_0\}$, where $\Psi_1, \Psi_2, ...$ are all 1–1 and θ_0 contains more than one element. (The functions Ψ_t and the set θ_0 are assumed known to the designer.)

II′. Special case: $\Psi_1, \Psi_2, ...$ are all identity mappings, so that $\rho_0 = \rho_1 = \rho_2 = \cdots$.

III. $\theta = \theta_0 \times \theta_1 \times \cdots$, where each θ_t contains more than one element.

III′. Special case: $\theta_0 = \theta_1 = \theta_2 = \cdots$, and the set θ_0 (hence also all sets θ_t, $t = 0, 1, ...$) is assumed known to the designer.

Thus if the multi-period economy is stationary and this fact is known to the designer then we are either in special case II′ (where the designer does not know the value of ρ_0) or I′ (where the designer does know ρ_0). In either case, the issue of informational decentralization is trivial.[6]

Furthermore, in situations of Type I (whether or not I′) the whole problem of designing a resource allocation mechanism becomes trivial because here the designer, knowing all the parameters of the problem, can directly calculate the desired (optimal[7] or efficient) actions (quantities to be consumed and/or produced) and instruct agents to carry them out. Clearly, the issue of informational decentralization does not arise;[8] more formally, such a resource allocation mechanism is privacy-preserving, since, in this case, the individual behavior rule (to carry out the prescribed action) does not require the knowledge of the others' parameters. (In fact, it does not require any knowledge on the part of the individual agents.)

Let us consider informational situations of Type II (whether or not II′). Here the designer can calculate an efficient (or optimal program) as a function of ρ_0, say $\langle c_t(\rho_0), x_t(\rho_0), y_t(\rho_0)\rangle$ where c, x, y denote respectively

[6] Provided the designer knows the initial stock $\bar{y}$, the discount factor δ and the felicity function $u_t(\cdot)$, $t \geqslant 0$ when optimality (rather than mere efficiency) is the goal. See Section IV.0 below for the optimality notion that we explore in detail.

[7] The comment in footnote 6 is also applicable here.

[8] Again, it should be stressed that we are assuming that the initial stock, the discount factor, and the felicity function(s) are known to the designer. Otherwise, the problem of desinging a decentralized mechanism even in a stationary environment ceases to be trivial.

consumption, input, and output. The individual agent t would then be instructed by the designer to choose the quantities

$$\tilde{c}_t(\rho_t) = c_t(\Psi_t^{-1}(\rho_0)),$$

$$\tilde{x}_t(\rho_t) = x_t(\Psi_t^{-1}(\rho_0)),$$

$$\tilde{y}_t(\rho_t) = y_t(\Psi_t^{-1}(\rho_0));$$

i.e., the designer supplies to agent t the functional relations[9] $\tilde{c}_t(\cdot)$, $\tilde{x}_t(\cdot)$, $\tilde{y}_t(\cdot)$, and agent t in turn finds their values for the armgument which, by hypothesis, is known to him. Again, the issue of informational decentralization is trivial.

It is natural, therefore, to eliminate from consideration situations such as those under I and II and to concentrate, as we shall, on situations of Type III.

Mechanisms such as those proposed above for situations of Type I and Type II are not, in general, "impersonal" in the sense of Hurwicz [2], hence from that point of view perhaps less desirable than the profit maximization process which is "impersonal." However, for certain stationary enviornments, it may be possible to obtain behavior rules that are independent of time, hence "impersonal," as well as privacy-preserving and that assure efficiency for infinite horizons, enough though the profit maximization mechanism does not (see Example 3.3). Hence profit maximization is not the only impersonal mechanism of interest.

A brief outline of the paper follows. Notational conventions are introduced in Sections II and IV.1.

Section III.1 describes basic aspects of multiperiod production economies, with Section III.2 focusing on the linear case. Section III.3 introduces certain general concepts in the theory of temporally decentralized resource allocation mechanisms, and provides illustrations involving the competitive programs, pure accumulation programs, programs with constant consumption/savings ratios, and single period consumption programs. It is then shown (Theorem T.3) that, provided the class of a priori admissible environments is sufficiently rich, it is impossible to design a temporally decentralized mechanism that is both non-wasteful and unbiased. It may be noted that the latter two concepts are based on the notion of efficiency rather than optimality, and do not involve the knowledge of preferences. Also, T.3 is valid for stationary economies as well as those with technology varying over time. By contrast, Sections IV and V postulate optimization in the sense of maximizing the discounted sum of felicities and time-varying (non-stationary) technologies. Here it is

[9] The designer is able to do this because the 1–1 functions Ψ_t are assumed known to him.

assumed that the designer does not know the future technologies and that the producer existing at time t lacks complete knowledge of future technologies; typically he/she knows current and possibly past technologies. The question to be answered is whether rules of behavior can be formulated which would be temporally decentralized and would assure that the resulting programs be optimal. It is shown that, even if preferences are assumed to be common knowledge, the design of such behavior rules is impossible. For the world of linear economies, this impossibility result is established in Theorem T.5.1 which shows that there is one "possibility" exception, when the felicity function is logarithmic. For a world including a sufficiently rich (in a specific sense) set of strictly concave functions, Theorem T.5.2 shows that it is impossible to design a temporally decentralized mechanism guaranteeing optimality.

In arriving at these results we use a notion of sensitivity of the optimality criterion to changes in technology. Theorem T.4 relates this notion of sensitivity to that of temporal decentralizability.

It should be noted that the class of mechanisms under consideration is restricted to those of the "proposed program" type, i.e., that do not explicitly use auxiliary variables such as prices. However, by allowing the decisions at a given time to be based on past as well as current technologies (and so relaxing the requirements of temporal decentralization), we broaden the class of allowable mechanisms sufficiently to include an equivalent of the competitive (price) mechanism. We have not, however, compared systematically the class of proposed program mechanisms permitting the utilization of past productivity data with the class of mechanisms using auxiliary variables but permitting the use of current productivity data only.

Finally, in Sedtion VI we consider a stronger notion of sensitivity. Whereas in T.5.2 it is shown that the initial optimal decisions vary as a result of a permanent change from one technology to another, we show in T.6.1 that they change in response even to a temporary (one period) change in technology. With a different class of admissible environments (technological futures) and even with a more liberal notion of temporal decentralization (the verification rule can be allowed to depend on all available information in an infinite but incomplete set of periods), we get the impossibility of designing decentralized mechanisms guaranteeing optimality.

II. Notation

The set of non-negative integers is denoted by $N = \{0, 1, 2, \ldots\}$. R (resp. P) denotes the set of reals (resp. positive reals). $\bar{P}$ is the set of nonnegative reals. For any two infinite sequences $\mathbf{x}=(x_t)$ and $\mathbf{x}'=(x_t')$ of real numbers, we write $\mathbf{x}\geqslant\mathbf{x}'$ if $x_t\geqslant x_t'$ for all t in N; we write $\mathbf{x}>\mathbf{x}'$ if $x\geqslant x'$ and $\mathbf{x}\neq\mathbf{x}'$; we write $\mathbf{x}\gg\mathbf{x}'$ if $x_t>x_t'$ for all t in N. We denote by $\mathbf{0}$ the infinite sequence with each term equal to 0; i.e., $\mathbf{0}=(0, 0, ...)$. A sequence $\mathbf{x}=(x_t)$ of real number is *non-negative* if $\mathbf{x}\geqslant 0$; $\mathbf{x}$ is *strictly positive* if $\mathbf{x}\gg\mathbf{0}$. The set of all non-negative real sequences is denoted by S^+ and the set of all strictly positive sequences is denoted by $\mathring{S}$. Given a sequence $\mathbf{x}$, the symbol $\mathbf{x}^{(T)}$ denotes the $(T+1)$-vector whose elements are the first $T+1$ elements of $\mathbf{x}$. The T-fold cartesian product of a set A with itself is denoted by A^T (T may be countably infinite, in which case we write $A^{(\infty)}$). For any positive integer $T\geqslant 1$, let X_T denote the space of all infinite sequences $\mathbf{x}=(x_t)$ such that $x_t>0$ for $t=0, ..., T$, and $x_t=0$ for $t\geqslant T+1$. We write $X=\bigcup_{T=1}^{\infty} X_T$.

III. A Model of Intertemporal Resource Allocation

Some of the central questions on designing a resource allocation mechanism that is informationally decentralized over time can be discussed formally in the context of a standard one-good model. Traditionally, such models have been useful in bringing into sharp focus some essential features of intertemporal economics. In our case, the model seems to provide the simplest framework for isolating those features of a dynamic economy that are critical in limiting the possibilities of decentralization of decision-making. The assumption that there is a single good in each period helps us to abstract away from the usual *intratemporal* allocation and decentralization problems and to concentrate solely on *intertemporal* aspects.

III.1. *The Economy over Time*

Suppose that there is a single producible good in the economy, the stock of which in period t is denoted by a non-negative number y_t. The *initial stock* is given by a positive number $\bar{y}$. The stock can be used either in consumption or as an input to generate the output of the same good in the next period. There is no exogenous supply after the initial period. (The model is "closed" in the sense of von Neumann). Hence, the stock in period $t+1$ is precisely the output of the good resulting from the input in period t. Let c_t and x_t denote the "consumption" and "input" (or "capital") in period t, respectively. The technological relationship between y_{t+1} and x_t is specified as

$$0\leqslant y_{t+1}=f_{t+1}(x_t) \qquad \text{for all } t\geqslant 0, \tag{3.1}$$

where $f_{t+1}: \bar{P} \to \bar{P}$ is the gross output function in period $t+1$. When we consider a finite horizon economy, $t = 0, \ldots, T$, while for an infinite horizon economy we let $t \in N$.

Let J be a class of functions $f: \bar{P} \to \bar{P}$ with the following properties:

(A.3.1) $f(0) = 0$.

(A.3.2) *f is increasing and continuous on $\bar{P}$.*

We shall refer to J as the set of all possible technologies or, equivalently, all possible gross output functions. Of particular interest is the case when J contains any of the following two well-studied classes ($\mathscr{L}$ and $\mathscr{F}$):

Class 1. Let $\mathscr{L}$ be the class of all *linear* homogeneous gross output functions of the form $f(x) \equiv \rho x$ where $\rho > 0$. We interpret ρ as the output–capital ratio.

Class 2. Let $\mathscr{F}$ be the class of all functions $f: \bar{P} \to \bar{P}$ that satisfy, in addition to (A.3.1) and (A.3.2), the following:

(A.3.3) *f is twice differentiable on P; $f'(x) > 0$, $f''(x) < 0$ for all $x \in P$; $\lim_{x \downarrow 0} f'(x) = \infty$ and $\lim_{x \uparrow \infty} f'(x) = 0$.*

(A.3.4) *$f(x)$ tends to infinity as x tends to infinity.*

We call $\mathscr{F}$ the class of all neoclassical gross output functions.[10] A parametric example of a neoclassical gross output function is $f(x) = x^{\alpha}$, $0 < \alpha < 1$.

The technological possibilities over time are described by a sequence $\mathbf{f} = (f_{t+1})$ of gross functions where $f_{t+1} \in J$ for all $t \geqslant 0$. *In what follows, we denote by* $\mathbf{f} = (f^{(\infty)})$ *the sequence having the property that* $f_{t+1} = f \in J$ *for all* $t \geqslant 0$. In other words, $(f^{(\infty)})$ *describes a stationary or time-invariant technology where the same gross output function f occurs in every period.* Given $\mathbf{f}$ and $\bar{y} > 0$, a feasible *program* of resource allocation *from* the initial stock $\bar{y} > 0$ consists of non-negative sequences $\mathbf{x} = (x_t)$, $\mathbf{y} = (y_t)$ and $\mathbf{c} = (c_t)$ which satisfy for all $t \in N$, the following:

$$y_0 = \bar{y}$$
$$c_t + x_t = y_t \tag{3.2}$$
$$0 \leqslant y_{t+1} \leqslant f_{t+1}(x_t).$$

The pair $(\mathbf{x}, \mathbf{y})$ is called a *feasible production program* and the sequence $\mathbf{c} = (c_t)$ is a *feasible consumption program* from $\bar{y}$. For brevity, we write "a

[10] $\mathscr{F}$ is not a subset of $\mathscr{L}$ because of the requirements in (A.3.3).

feasible $(\mathbf{x}, \mathbf{y}, \mathbf{c})_{\bar{y}}$" to denote a feasible program of resource allocation from the initial stock $\bar{y}$ (given the sequence $\mathbf{f}$)[11].

A program is evaluated according to criteria that are defined exclusively in terms of its sequence of consumptions. We shall study two such criteria: first, intertemporal efficiency, then a particular type of optimality. A feasible $(\mathbf{x}, \mathbf{y}, \mathbf{c})_{\bar{y}}$ is *intertemporally efficient* if there is no other feasible $(\mathbf{x}', \mathbf{y}', \mathbf{c}')_{\bar{y}}$ such that $\mathbf{c}' > \mathbf{c}$.

EXAMPLE 3.1 (Pure accumulation programs). Consider the feasible $(\hat{\mathbf{x}}, \hat{\mathbf{y}}, \hat{\mathbf{c}})_{\bar{y}}$ defined by:

$$\hat{x}_t = \hat{y}_t; \quad \hat{c}_t = 0; \quad \hat{y}_{t+1} = f_{t+1}(\hat{x}_t) \qquad \text{for all } t \geqslant 0. \tag{3.3}$$

We refer to this program as the program of pure accumulation. It is easy to check that from any $\bar{y} > 0$ this program is *not* efficient.

EXAMPLE 3.2 (Single period consumptions programs). Consider the feasible $(\mathbf{x}^0, \mathbf{y}^0, \mathbf{c}^0)_{\bar{y}}$ defined by:

$$\begin{aligned} c_0^0 = y_0^0 = \bar{y}; \qquad & x_0^0 = 0 \\ c_t = y_t = 0 \qquad & \text{for all } t \geqslant 1. \end{aligned} \tag{3.4}$$

This is the program in which the entire initial stock is consumed in period zero. Consumptions, inputs, and stocks are all zero in *all* periods $t \geqslant 1$. $(\mathbf{x}^0, \mathbf{y}^0, \mathbf{c}^0)_{\bar{y}}$ *is* intertemporally efficient. For an infinite-horizon model an infinite number of similar efficient programs can be constructed as follows. For any integer $T \geqslant 1$, define a feasible $(\mathbf{x}^T, \mathbf{y}^T, \mathbf{c}^T)$ by

$$\begin{aligned} & x_0^T = y_0^T = \bar{y} \\ & x_t^T = y_t^T; \quad c_t^T = 0 \qquad && \text{for} \quad t = 0, \ldots, T-1 \\ & c_T^T = y_T^T; \quad x_T^T = 0 \\ & c_t^T = y_t^T = x_t^T = 0 \qquad && \text{for} \quad t \geqslant T+1 \\ & y_{t+1}^T = f_{t+1}(x_t^T) \qquad && \text{for} \quad t \geqslant 0. \end{aligned} \tag{3.5}$$

In all periods 0 through $T-1$ the feasible $(\mathbf{x}^T, \mathbf{y}^T, \mathbf{c}^T)_{\bar{y}}$ prescribes zero con-

[11] Note that if we consider a finite-horizon model where the technologies over the T periods are described by $(f_{t+1})_{t=0}^{T-1}$, we can define a corresponding sequence as $\mathbf{f} = ((f_{t+1})_{t=0}^{T-1}, n^{(\infty)})$ where n: $\bar{P} \to \bar{P}$ is the function $n(x) \equiv 0$, for all $x \in \bar{P}$. Similarly, given a finite-horizon program $\langle (x_t)_{t=0}^T, (y_t)_{t=0}^T, (c_t)_{t=0}^T \rangle$ we can define a corresponding infinite sequences $\mathbf{x} = ((x_t)_{t=0}^T, \mathbf{0})$, $\mathbf{y} = ((y_t)_{t=0}^T, \mathbf{0})$, $\mathbf{c} = ((c_t)_{t=0}^T, \mathbf{0})$, as elements of S^+. Using such constructions we shall treat finite-horizon models as special cases of infinite-horizon models.

sumption, and requires that the entire stock be used as input. In period T, the entire stock is consumed, and there is no input. In all subsequent periods $t \geqslant T+1$, consumptions, inputs, and stocks are all zero.

III.2. *The Linear Model*

We shall now state some particularly sharp results on feasibility and efficiency in the case when all the gross output functions $f_{t+1}(x)$ are in $\mathscr{L}$; i.e.,

$$f_{t+1}(x) \equiv \rho_{t+1} x, \qquad \text{where} \quad \rho_{t+1} > 0, \qquad \text{for} \quad t \geqslant 0. \tag{3.6}$$

We shall use the sequence $\boldsymbol{\rho} = (\rho_{t+1})$ of output–capital ratios to represent the sequence of technologies. We require that the sequence $\boldsymbol{\rho}$ be either in $\mathring{S}$ (which corresponds to the case where production is possible over an infinite number of periods) or in X (which means that there cannot be any production after a finite number of periods). We use the notation

$$\sigma_0 \equiv 1, \quad \sigma_t = \prod_{i=1}^{t} \rho_i \qquad \text{for} \quad t \geqslant 1. \tag{3.7}$$

For a finite-horizon economy $\boldsymbol{\rho}$ is in X i.e., for some finite T, $\sigma_t = 0$ for $t > T$. In this case we interpret the notation $\sum_{t=0}^{\infty} c_t/\sigma_t$ as the finite sum $\sum_{t=0}^{T} c_t/\sigma_t$.

We state the following result characterizing feasible and efficient programs in a linear model:

R.3.1. *If* $\boldsymbol{\rho}$ *is in* $\mathring{S}$ *or in* $\bar{X}$, *and* y *is in* P, *then for any feasible* $(\mathbf{x}, \mathbf{y}, \mathbf{c})_{\bar{y}}$,

$$\sum_{t=0}^{\infty} [c_t/\sigma_t] \leqslant \bar{y}. \tag{3.8}$$

A feasible $(\mathbf{x}, \mathbf{y}, \mathbf{c})_{\bar{y}}$ *is efficient if and only if*

$$\sum_{t=0}^{\infty} [c_t/\sigma_t] = \bar{y}. \tag{3.9}$$

The proof of R.3.1 can be obtained as a special case of the proof in Majumdar [8, Theorem 4.1, pp. 364–368]. Using R.3.1 it is easy to verify that the program of pure accumulation defined in Example 3.1 is not efficient, whereas the programs $(\mathbf{x}^T, \mathbf{y}^T, \mathbf{c}^T)_{\bar{y}}$ defined in Example 3.2 are indeed efficient. We shall consider another class of intertemporally efficient programs with some interesting features:

EXAMPLE 3.3 (Programs with constant consumption/savings ratio). Consider an infinite-horizon economy. Let $\boldsymbol{\rho}$ be in $\mathring{S}$, and θ in $(0, 1)$. Consider the feasible $(\mathbf{x}^*, \mathbf{y}^*, \mathbf{c}^*)_{\bar{y}}$ defined by

$$y_0^* = \bar{y}$$

$$\left.\begin{array}{l} c_t^* = \theta y_t^*, \quad x_t^* = (1-\theta)\, y_t^* \\ y_{t+1}^* = \rho_{t+1} x_t^* \end{array}\right\} \text{for all } t \in N. \tag{3.10}$$

By R.3.1, $(\mathbf{x}^*, \mathbf{y}^*, \mathbf{c}^*)_{\bar{y}}$ is intertemporally efficient. Observe that (a) the decision rule regarding consumption (and input) is the same for all periods (since θ does not depend on time), and (b) the implementation of the decisions does not require any information about $\boldsymbol{\rho}$. At the beginning of any period $t \geqslant 0$, the total stock y_t^* is known, and the decision on consuming a fraction θ of the total stock can be carried out without having to remember past output–capital ratios or to anticipate future output–capital ratios. These features will be commented on later.

III.3. *Resource Allocation Mechanisms*

In the context of intertemporal efficiency problems the environment is defined by the initial stock $\bar{y}$ and the technological possibilities over time specified by the sequence $\mathbf{f} = (f_{t+1})_{t=0}^{\infty}$ of gross output functions. We denote such an environment by e and write $e = (\mathbf{f}, \bar{y})$.[12] Classes of environments are denoted by E with or without affixes. In particular, E^0 is the class of environments considered as conceivable in a given model.

Now let Z denote the class of conceivable programs. (In our paper, we may take $Z = S^+ \times S^+ \times S^+$.) The set of programs feasible in environment e (a subset of Z) will be written $Z^*(e)$, and Z^*: $E^0 \twoheadrightarrow Z$ is the feasibility correspondence.

The various evaluation criteria such as efficiency or optimality are also interpreted as correspondence from E^0 into Z; in fact, they are sub-correspondences of Z^*.

In particular, we shall denote by $Z^{**}(e)$ the set of programs intertemporally efficient for the environment e. Of course, $Z^{**}(e) \subset Z^*(e)$ since efficiency implies feasibility; Z^{**}: $E^0 \twoheadrightarrow Z$ is the intertemporal efficiency correspondence.

More generally, let $\mathcal{O}: E^0 \twoheadrightarrow Z$ be an evaluation criterion,[13] a sub-

[12] In optimality models the environment also includes information about performances and the discount factor. Thus $e = (\delta, \mathbf{u}, \bar{y}, \mathbf{f})$ where δ is the discount factor and $\mathbf{u} = (u_t, (\cdot))_{t=0}^{\infty}$ with $u_t(\cdot)$ the tth felicity function. In Section IV, for simplcity we consider only a stationary sequence of felicities $u_t(\cdot) = u(\cdot)$ for all $t \geqslant 0$.

[13] Also called social choice correspondence (rule) or performance correspondence in much of the literature.

correspondence of Z^*; i.e., $\mathcal{O}(e) \subset Z^*(e)$ for all e in E^0. An environment $e \in E^0$ is said to be *admissible for* $\mathcal{O}$ if the set $\mathcal{O}(e)$ of programs selected by the criterion $\mathcal{O}$, when e prevails, is non-empty. We denote by $E^0(\mathcal{O})$ the set of all environments admissible for $\mathcal{O}$. Example 3.2 shows that, for the intertemporal efficiency criterion Z^{**}, the set of all admissible environments $E^0(Z^{**})$ is $J^{(\infty)} \times P$. To lighten notation, we often write $E^0(Z^{**}) = E$. A generic element e of E is of the form $e = (\mathbf{f}, \bar{y})$.

Our object is to investigate resource allocation mechanisms that generate outcomes satisfying given evaluation criteria.

In general, a *resource allocation mechanism* for (E^0, Z) is defined as a triple $\pi = (\mathcal{M}, \mu, h)$ where $\mathcal{M}$ is a set called *the message space*, μ: $E^0 \twoheadrightarrow \mathcal{M}$ is the *equilibrium correspondence* and h: $\mathcal{M} \to Z$ the *outcome function.*[14]

We say that a mechanism $\pi = (\mathcal{M}, \mu, h)$ *realizes*[15] an evaluation criterion $\mathcal{O}$: $E^0 \twoheadrightarrow Z$ *on* $E' \subset E$ if and only if

(1) $\mu(e) \neq \varnothing$ for all $e \in E'$;

and

(2) "$m \in \mu(e)$ and $z = h(m)$" implies "$z \in \mathcal{O}(e)$".

When property (2) of the preceding definition holds, we say that π is *non-wasteful* for $\mathcal{O}$ on E'. Furthermore, π is said to be *unbiased* for $\mathcal{O}$ on E' if it is the case that

(3) for every $e \in E'$ and every $z \in \mathcal{O}(e)$ there is a message $m \in \mathcal{M}$ such that $m \in \mu(e)$ and $z = h(m)$.

Finally, we say that π is *satisfactory* for $\mathcal{O}$ on E' if it is both unbiased and non-wasteful for $\mathcal{O}$ on E'. If the evaluation criterion is not explicitly mentioned, it is understood to be intertemporal efficiency.

If $m \in \mu(e)$ and $z = h(m)$, we say that z is an *equilibrium program* for e. π is *non-wasteful* if all equilibrium programs are non-wasteful; it is *unbiased* if every program satisfying the evaluation criterion is an equilibrium program.

Consider now an environment e whose description can be factored into characteristics of individual participants. Let the set of participants (finite or countably infinite) be denoted by $I = \{0, 1, 2, \ldots\}$. We may then write $e = (e^{-1}, e^0, e^1, \ldots)$ where e^i, $i \geqslant 0$, is the characteristic of the ith participant, and e^{-1} contains information (common knowledge) not attributable to any individual participant. We denote by E^{0i} the class of conceivable charac-

[14] For more details, see Mount and Reiter [13] and Hurwicz [4].

[15] Occasionally authors use "implements" instead of "realizes," but this creates confusion with concepts involving game theoretic equilibria, e.g., where $\mathcal{M}$ is the product of individual strategy sets and $\mu(e)$ is the set of Nash equilibria for e.

teristics of individual $i \in I$. We have $E = E^{0,-1}(\prod_{i \in I} E^{0i})$ where $E^{0,-1}$ is the range of conceivable values of e^{-1}.

A resource allocation mechanism is said to be *privacy-preserving*[16] if there exists for each individual $i \in I$ *individual equilibrium correspondence* μ^i: $E^{0i} \twoheadrightarrow \mathcal{M}$ such that for any $e = (e^0, e^1, e^2, \ldots)$ in E^0, $\mu(e^0, e^1, e^2, \ldots) = \bigcap_{i \in I} \mu^i(e^i)$.

EXAMPLE 3.4. A large part of the literature on infinite-horizon economies has stressed that a "competitive" or "intertemporal profit maximizing" mechanism is *not* non-wasteful. A feasible program $(\mathbf{x}', \mathbf{y}', \mathbf{c}')_{\bar{y}}$ is *competitive* if there is some $\mathbf{p} = (p_t)$ in $\mathring{S}$ such that

$$p_{t+1} f_{t+1}(x'_t) - p_t x'_t \geqslant p_{t+1} y - p_t x, \qquad \text{for all } (x, y)$$
$$\text{satisfying} \quad 0 \leqslant y \leqslant f_{t+1}(x), \qquad x \geqslant 0. \tag{3.11}$$

The sequence $\mathbf{p}$ is referred to as the sequence of *competitive prices* and (3.11) is the condition of *intertemporal profit maximization.*

We shall now define the *competitive resource allocation mechanism* applicable to this example.

Here the generic element m of the message space $\mathcal{M}$ is of the form $m = (\mathbf{z}, \mathbf{p})$ where $\mathbf{z} = (\mathbf{x}, \mathbf{y}, \mathbf{c}) \in Z \equiv \mathring{S} \times \mathring{S} \times \mathring{S}$, and $\mathbf{p} \in \mathring{S}$, so that $\mathcal{M} = Z \times \mathring{S}$, $Z = \mathring{S} \times \mathring{S} \times \mathring{S}$.

The set $I = \{0, 1, \ldots\}$ of participants is indexed by the time subscript t. The environment is factorable so that $E^0 = \prod_{t=0}^{\infty} E^{0t}$, where the generic element of E^{0t} is f_{t+1} for $t \geqslant 1$ and $(f_1, \bar{y})$ for E^{00} (to simplify matters we omit the common knowledge component $E^{0,-1}$).

The mechanism is privacy-preserving. The tth equilibrium correspondence μ^t is given, for $m' = ((\mathbf{x}', \mathbf{y}', \mathbf{c}'), \mathbf{p})$, by

$$m' \in \mu^t(e^t) \Leftrightarrow \begin{cases} y'_{t+1} = f_{t+1}(x'_t) \quad \text{for } t \geqslant 0, \; y'_0 = \bar{y}; \quad \text{and} \\ p_{t+1} f_{t+1}(x'_t) - p_t x_t \geqslant p_{t+1} y - p_t x \\ \text{for all } (x, y) \text{ satisfying } 0 \leqslant y \leqslant f_{t+1}(x), \quad x \geqslant 0, \quad t \in N; \text{ and} \\ c'_t = y'_t - x'_t \qquad \text{for all } t \geqslant 0. \end{cases}$$

Finally, the outcome function h is given, for $m' = ((\mathbf{x}', \mathbf{y}', \mathbf{c}'), \mathbf{p})$, by

$$h(m') = (\mathbf{x}', \mathbf{y}', \mathbf{c}').$$

That is, it is a projection on the program component space.

[16] Sometimes the term *informationally decentralized* is used instead of *privacy-preserving*. But in Hurwicz [2] the term "informationally decentralized" has a narrower meaning.

The program of *pure accumulation* $(\hat{\mathbf{x}}, \hat{\mathbf{y}}, \hat{\mathbf{c}})_{\bar{y}}$ defined in Example 3.1 can be shown to be competitive if one defines $\mathbf{p} = (p_t)$ by

$$p_0 \equiv 1, \quad p_{t+1} = p_t / f'_t(\hat{x}_t) \qquad \text{for} \quad t \geqslant 0. \tag{3.12}$$

This program is not efficient. Hence, the competitive mechanism is *not non-wasteful*, on the set E for the efficiency criterion Z^{**}. On the other hand, the fundamental result of Malinvaud [10] can be directly used to prove that for any $(\mathbf{f}, \bar{y})$ in $J^{(\infty)} \times P$, an intertemporally efficient program is competitive. Thus, the competitive resource allocation mechanism is *unbiased* on E for the efficiency criterion. That this is so is particularly easy to see in the linear model. Here it is simple to verify directly that if $(\mathbf{x}', \mathbf{y}', \mathbf{c}')_{\bar{y}}$ is efficient in the environment $(\boldsymbol{\rho}, \bar{y}) \in \mathring{S} \times P$, it si *competitive* relative to the price system $\mathbf{p} = (p_t)$ defined by

$$p_0 = 1 = 1/\sigma_0, \qquad p_t = 1/\sigma_t \qquad \text{for} \quad t \geqslant 1. \tag{3.13}$$

Note that the prices p_t depend only on $(\boldsymbol{\rho}^{(t)})$ and do not require any knowledge of the environment at date $t+1$ onwards.

Example 3.3 (Constant consumption/savings ratio (cont.)). We now go back to Example 3.3 and reconsider the feasible $(\mathbf{x}^*, \mathbf{y}^*, \mathbf{c}^*)_{\bar{y}}$ which can also be defined as

$$\left.\begin{aligned} y_t^* &= (1-\theta)^t \sigma_t \bar{y} \\ x_t^* &= (1-\theta)^{t+1} \sigma_t \bar{y} \\ c_t^* &= \theta(1-\theta)^t \sigma_t \bar{y} \end{aligned}\right\} \text{for all } t \geqslant 0. \tag{3.14}$$

Here we define the following resource allocation mechanism $\pi^* = (\mathcal{M}^*, \mu^*, h^*)$. Let $\mathcal{M}^* = Z = \mathring{S} \times \mathring{S} \times \mathring{S}$. Let $\mu^*(e) = \bigcap_{t=0}^{\infty} \mu^*(e^t)$, with

$$\left.\begin{aligned} &m^* = (\mathbf{x}^*, \mathbf{y}^*, \mathbf{c}^*), \\ &m^* \in \mu^{*t}(e^t) \end{aligned}\right\} \Leftrightarrow \left\{\begin{aligned} &y_t^* = f_t(x_{t-1}^*) \quad \text{for} \quad t \geqslant 1 \quad \text{and} \quad y_0^* = \bar{y} \\ &x_t^* = (1-\theta)\, y_t^* \quad \text{and} \\ &c_t^* = \theta y_t^* \quad \text{for} \quad t \geqslant 0. \end{aligned}\right. \tag{3.15}$$

Finally, let h be the identity function; i.e., $h(\mathbf{x}^*, \mathbf{y}^*, \mathbf{c}^*) = (\mathbf{x}^*, \mathbf{y}^*, \mathbf{c}^*)$.

Clearly, this mechanism realizes the constant consumption/savings ratio program, and is non-wasteful—but not unbiased—for the intertemporal efficiency criterion. It is privacy-preserving, and the knowledge of the environment is only needed to verify feasibility. As for the parameter θ, it is assumed to be common knowledge.

For any sequence $\boldsymbol{\rho} = (\rho_t)$ in $\mathring{S}$, we refer to $\rho^{(\tau)} \equiv (\rho_1, ..., \rho_\tau)$ as a *partial history* up to period τ. Note that (3.15) expresses the variables in period t,

(x_t^*, y_t^*, c_t^*) as functions of the initial stock $\bar{y}$ and the partial history up to period t. Given the special structure of the model, the relationships (3.14) are given in terms of particularly simple functional forms. Conceptually, the example suggests a formal definition of an *informationally (temporally) decentralized resource allocation mechanism.*

However, the privacy-preserving property introduced above does not adequately express the intuitive notion of informational decentralization in intertemporal models. We think of the relation $m \in \mu^t(e^t)$ as one verified by an individual existing at time t. Now in intertemporal models m itself is an infinite sequence of the form $\mathbf{m} = (m_t)_{t=0}^{\infty}$, $m_t \in \mathcal{M}^t$, $\mathcal{M} = \prod_{t=0}^{\infty} \mathcal{M}^t$. (For instance, in the competitive mechanism $\mathbf{m} = (\mathbf{x}, \mathbf{y}, \mathbf{c}, \mathbf{p})$. In the constant consumption/savings ratio model $\mathbf{m} = (\mathbf{x}, \mathbf{y}, \mathbf{c})$.) Typically, the individual existing at time τ will not be able to observe the values of m_t for $t > \tau$. Hence it is more natural to require that, for each t, μ^t be such that, for $\mathbf{m} = (m_t)_{t=0}^{\infty}$, the relation $m_t \in \mu^t(e^t)$ be equivalent to $(m_\tau)_{\tau=0}^{t} \in \nu^t(e^t)$ for some correspondence $\nu^t \colon E^{0t} \twoheadrightarrow \prod_{\tau=0}^{t} \mathcal{M}^\tau$.

The interpretation of this requirement is that the individual existing at time t need not look at future messages $(m_{t+1}, \dots)$ to verify whether the proposed infinite sequence $\mathbf{m} = (m_0, m_1, \dots, m_t, m_{t+1}, \dots)$ is an acceptable candidate for equilibrium from his/her viewpoint. A sequence $\mathbf{m}$ is an equilibrium sequence if it is acceptable to individuals existing at all timepoints $t \geqslant 0$. For this verification she/he, at most, needs to know his/her own characteristic e together with the past and present messages, i.e., $m_0, m_1, \dots, m_t$.

If, however, she/he were actually required to know the messages all the way back to m_0, the individual's capacity for "remembering" messages would have to grow without bounds as t tends to infinity. To avoid this, it is not unreasonable to impose a uniform (with respect to t) bound on the memory requirements. For the sake of simplicity we choose the lowest bound requiring only the knowledge of current m but no memory of $m_{t-1}, m_{t-2}, \dots$. We thus have $\nu^t \colon E^{0t} \twoheadrightarrow \mathcal{M}^t$.

In a similar spirit we require that the outcome function $h = (\tilde{h}^0, \tilde{h}^1, \dots)$, so that $h(m) = (\tilde{h}^0(m), \tilde{h}^1(m), \dots)$, with $\tilde{h}^t(m)$ depending on the current value m_t only; i.e., $\tilde{h}^t(m) = h^t(m_t)$ and $h(m) = (h^0(m_0), h^1(m_1), \dots)$. Thus $z_t = h^t(m_t)$; $z_t = (x_t, y_t, c_t)$.

The interpretation of the process as now defined is as follows. Think of $\mathbf{m} = (m_0, m_1, \dots)$ as an infinite message sequence being proposed to the (infinite) collectivity of all individuals who will exist respectively at time $t = 0, 1, \dots$. However, the individual existing at time t only sees the component m_t of m and checks whether it is compatible with his/her characteristic, i.e., whether $m_t \in \nu^t(e^t)$. If it is, the corresponding outcome value is calculated as $z_t = h_t(m_t)$. If all individuals say "yes," i.e., if $m_t \in \nu^t(e^t)$ for all $t \geqslant 0$, then $\mathbf{m}$ is an equilibrium message sequence and $\mathbf{z} = (z_0, z_1, \dots)$ is an equilibrium program. If even one individual says "no," i.e., if $m_{t'} \notin v^{t'}(e^{t'})$ for some $t' \geqslant 0$, then $\mathbf{m}$ is not an equilibrium message sequence and some other message sequence, say $\mathbf{m}' \neq \mathbf{m}$, must be proposed as a candidate for an equilibrium message sequence.

In the present paper the preceding scenario is modified in several aspects. First, consider the message space $\mathscr{M}$. In many economic models it is natural to think of the generic message m as being of the form $m = (a, v)$ where a is a proposed action (e.g., resource allocation) and v some auxiliary (usually multidimensional) variable. Thus, $\mathscr{M} = Z \times V$. For instance, in the competitive mechanism above we had $\mathbf{m} = (\mathbf{z}, \mathbf{p})$ where $\mathbf{z}$ is a proposed program and $\mathbf{p}$ a price sequence. Furthermore, the outcome function h is the projection on the component space; i.e., $h(\mathbf{m}) \equiv h(a, v) \equiv a$. Thus, in the competitive example, $h(\mathbf{m}) \equiv h(\mathbf{z}, \mathbf{p}) \equiv \mathbf{z}$.

A special subclass of such mechanisms is one in which no auxiliary variable enters, so that $\mathscr{M} = Z$ and $h(m) \equiv h(z) \equiv z$; i.e., the message space is the action space (in our models, the program space), and the outcome function h is the identity function. The interpretation is as follows. An infinite program $\mathbf{z} = (z_0, z_1, ...)$ is proposed for consideration of the (infinite) collectivity of all individuals. However, the individual existing at time t sees only the proposed value of z_t and verifies whether it is compatible with his/her characteristic, i.e., whether $z_t \in v^t(e^t)$, where the correspondence v^t is the rule being applied in the verification. If all individuals answer "yes," i.e., if $z_t \in v^t(e^t)$ for all $t \geqslant 0$, then $\mathbf{z}$ is an equilibrium program, otherwise it is not. This type of mechanism has been called a "proposed action process" (Hurwicz [3]).

It is of interest to know whether a given evaluation criterion $\mathscr{O}$ can be realized by a proposed action process. For instance, as seen above, the constant consumption/savings ratio criterion can be so realized because the mechanism that realizes it (as described above) is of the proposed action type. ("Proposed program" is actually more appropriate here.)

On the other hand, the competitive mechanism described above is not of the "proposed program" type because prices are introduced as auxiliary variables. Since we do not want to exclude competitive and similar mechanisms from consideration, but at the same time want to avoid introducing auxiliary variables, we formally relax the privacy-preserving requirement by permitting the individual existing at time t to know not only his/her own characteristic, but also those of his predecessors, i.e., the value of $e^0, e^1, ..., e^{t-1}$.[17] Thus the verification rule may be written as $z \in v^t(e^0, e^1, ..., e^{t-1}, e^t)$. We note that a reinterpretation of the competitive

[17] In our examples, $e^t = (f_{t+1}, \bar{y})$; i.e., the individual existing at time t knows the production function f_{t+1} which she/he is using.

mechanism in linear environments fits this formulation. We need only to replace p_t in (3.11) by $1/\prod_{\tau=0}^{t} \sigma_\tau$ for all $t \geqslant 0$ (here $\sigma_0 = 1$ and $\sigma_t = \prod_{\tau=1}^{t} \rho_\tau$ for $t \geqslant 1$).

To be sure, this amounts to introducing, at least to some extent, auxiliary variables by the back door.[18] But, as discussed in the Introduction, relaxation of the privacy-preserving requirement only strengthens the *impossibility* results to be presented below.

One more minor relaxation of requirements which, again, does no harm in impossibility results. Instead of only permitting that the individual existing at time t know the characteristics of the predecessor $(e^0, e^1, ..., e^{t-1})$ in addition to his/her own (e^t), we define a sequence of positive integers, say $T_1, T_2, ..., T_t, ...,$ for all $t \geqslant 0$ such that the individual existing at time t may be permitted to know the characteristics up to and including e^{T_t}. If $T_t > t$, this means that the individual existing at time t may know the characteristics of his/her successors as well as predecessors. Formally, the correspondence representing the (*temporally*) *decentralized verification rule* will be denoted by ψ_t, and the verification rule becomes $z_t \in \psi_t(e^0, ..., e^{T_t})$. In Example 3.3, the sequence $\{T_t\}$ is simply $\{t\}$. In general, we allow for the possibility that the decisions in any period depend on a *finite* number of periods in the future also (i.e., it may be that $T_t > t$).

To relate our present model of an intertemporal resource allocation mechanism to the definition of $\pi = (\mathcal{M}, \mu, h)$ (see Section III.3), we have (1) $\mathcal{M} = Z$ where Z is the space of conceivable programs, (2) μ is defined by the sequence $(\psi_t)_{t=0}^{\infty}$ of decentralized verification rules, and (3) the outcome function $h: Z \to Z$ is a sequence of projections, $h = (h^0, ..., h^t, ...)$ where $h^t(\mathbf{z}) = z_t$. Our discussion of Example 3.3 can be summarized as follows: *in the linear model, even when the designer does not know the true environment, one can construct a resource allocation mechanism that is informationally temporally decentralized and non-wasteful.* (Recall, however, the mechanism of Example 3.3 is *not* unbiased). The impossibility of designing an informationally (temporally) decentralized resource allocation mechanism that is *both* non-wasteful and unbiased follows from:

T.3. *Let* $E = J^{(\infty)} \times P$, and $(\mathbf{f}, \bar{y}) \in E$.[19] *There is no pair of sequences* (T_t) *and* (Ψ_t) *such that* (i) (T_t) *is a sequence of finite positive integers and* (ii) (Ψ_t) *is a sequence of correspondences with* $\Psi_t\colon (\mathbf{f}^{(T_t)}, \bar{y}) \to \bar{P}^3$, *and such that* $(\mathbf{x}, \mathbf{y}, \mathbf{c})_{\bar{y}}$ *is in* $Z^{**}(\mathbf{f}, \bar{y})$ *if and only if* $(x_t, y_t, c_t) \in \Psi_t$ *for all* $t \geqslant 0$.

[18] It has also the disadvantage of requiring a memory capacity that grows without bounds as t tends to infinity. However, as in the competitive example, it may have an equivalent *with* auxiliary variables in which memory requirements are small or uniformly bounded with respect to t.

[19] Hence, by Example 3.2, there is at least one program for $(\mathbf{f}, \bar{y})$.

Proof. See (A.1) in the Appendix.

Remark. The proof of T.3 does not use any property of the sequence $\mathbf{f} = (f_{t+1})$ of gross output functions. Thus, one can restrict or expand the class of admissible technologies and continue to get the same result. In particular, T.3 is valid even for a stationary linear economy (where $f_{t+1}(x) = \rho x$ for all $t \geqslant 0$).

IV. Intertemporal Optimality and Decentralization

IV.0. *Optimality*

The proof of T.3. does exploit the multiplicity of intertemporally efficient programs in a particular environment. The "negative" nature of T.3. leads one to speculate whether one can derive more "positive" results—on the "possibility" rather than "impossibility" of intertemporal decentralization—when the evaluation criterion selects a unique feasible allocation in every admissible environment. We now turn to this question. The evaluation criterion that we explore (maximizing the "discounted" sum of one period utilities or felicities) has been particularly well studied in dynamic economics (see, e.g., Dasgupta and Heal [1] for a discussion of alternative motivations behind the use of such a criterion).

We continue to use the same definition of feasibility as before (see 3.2). We use a discount factor δ, $0 < \delta < 1$ and consider it as a part of the description of an environment. Let u be the one period utility or "felicity" function defined either on P or $\bar{P}$ with values in R. We interpret $u(c)$ to be a measure of satisfaction derived from consumption c. The following assumption on u is maintained throughout:

(A.4.1) *u is twice differentiable with $u'(c) > 0$, $u''(c) < 0$ at $c > 0$; when u is defined only over P, $\lim_{c \downarrow 0} u(c) = -\infty$. Also $\lim_{c \downarrow 0} u'(c) = \infty$ and $\lim_{c \uparrow \infty} u'(c) = 0$.*

Parametric examples of $u(c)$ are $u(c) = \log c$ and $u(c) = c^{\alpha}$, $0 < \alpha < 1$.

As we allow for an infinite number of planning periods, some consistency conditions on the "parameters" (the technology and the discount factor) are needed to make an optimization exercise meaningful. To begin with, consider a finite-horizon problem in whih one is given a finite sequence $(f_1, ..., f_T)$ of technologies, T being a finite positive integer, and $f_t \in J$. Given the initial stock $\bar{y}$, a *feasible* program $(\mathbf{x}, \mathbf{y}, \mathbf{c}) = \langle (x_t)_{t=0}^{T}, (y_t)_{t=0}^{T}, (c_t)_{t=0}^{T} \rangle$ from $\bar{y}$ must satisfy (see (3.2))

$$x_0 + c_0 = y_0 \equiv \bar{y}$$

$$x_t + c_t = y_t \qquad \text{for} \quad t = 1, \ldots, T$$

$$y_{t+1} \leqslant f_{t+1}(x_t) \qquad \text{for} \quad t = 0, \ldots, T-1$$

$$x_t \geqslant 0, \quad c_t \geqslant 0 \qquad \text{for} \quad t = 0, \ldots, T.$$

A program $(\mathbf{x}^*, \mathbf{y}^*, \mathbf{c}^*)$ from $\bar{y}$ is *optimal* (given $e \equiv ((f_t)_{t=1}^{T}, \bar{y}, \delta)$) if it is feasible and

$$\sum_{t=0}^{T} \delta^t u(c_t^*) \geqslant \sum_{t=0}^{T} \delta^t u(c_t) \tag{4.1}$$

for all feasible $(\mathbf{x}, \mathbf{y}, \mathbf{c})$ from the same $\bar{y}$. We say that $e = ((f_t)_{t=1}^{T}, \bar{y}, \delta, u)$ is *admissible* (for the optimality criterion (4.1)) if there is an optimal program given e.

We now extend the definition to the case of an infinite horizon. Given the initial stock $\bar{y} > 0$, an infinite sequence $\mathbf{f} = (f_{t+1})$ of technologies (where $f_{t+1} \in J$, the set of all possible gross output functions) and a discount factor $\delta \in (0, 1)$, if $(\mathbf{x}, \mathbf{y}, \mathbf{c})_{\bar{y}}$ is a feasible (infinite) program from $\bar{y}$ (satisfying (3.2)) we write

$$\bigcup [(\mathbf{x}, \mathbf{y}, \mathbf{c})_{\bar{y}}] = \sum_{t=0}^{\infty} \delta^t u(c_t) \tag{4.2}$$

whenever the right side is finite or minus infinity. A feasible infinite program $(\mathbf{x}^*, \mathbf{y}^*, \mathbf{c}^*)_{\bar{y}}$ is optimal if $\bigcup (\mathbf{x}^*, \mathbf{y}^*, \mathbf{c}^*)$ is finite and for all feasible $(\mathbf{x}, \mathbf{y}, \mathbf{c})_{\bar{y}}$,

$$\bigcup [(\mathbf{x}^*, \mathbf{y}^*, \mathbf{c}^*)_{\bar{y}}] \geqslant \bigcup [(\mathbf{x}, \mathbf{y}, \mathbf{c})_{\bar{y}}]. \tag{4.3}$$

We continue to refer to an environment $e = (\mathbf{f}, \bar{\mathbf{y}}, \delta, u)$ as *admissible* for the optimality criterion (4.3), if there is an optimal program in e. Let E be the set of all admissible environments. E is non-empty.

Remark. Clearly, the set E contains all *finite*-horizon environments $e = ((f_t)_{t=1}^{T}, \bar{y}, \delta, u)$ such that $f_t \in J$ for all $t = 1, \ldots, T$, $y > 0$, $0 < \delta < 1$.

Next, note that if f is "neo-classical" (more precisely, if f satisfies (A.3.1) through (A.3.4)), and the technological possibilities are described by a stationary sequence $\mathbf{f} = (f^{(\infty)})$, then $e = (\mathbf{f}, y, \delta, u)$ is admissible for all $y > 0$, $\delta \in (0, 1)$ (use, e.g., Theorem 3.1 of Majumdar [9]).

As a third example, suppose $u(c) = c^\alpha/\alpha$, $0 < \alpha < 1$, and the set of all admissible technologies consists exclusively of all linear functions $f(x) = \rho x$, $\rho > 0$ (i.e., $J = \mathscr{L}$). Corresponding to any $\bar{y} > 0$ and $\boldsymbol{\rho} = (\rho_{t+1}) \in \mathring{S}$, consider the program of pure accumulation $(\hat{\mathbf{x}}, \hat{\mathbf{y}}, \hat{\mathbf{c}})_{\bar{y}}$ defined in (3.3) (here $\hat{y}_t = \sigma_t \bar{y}$). One can prove that if $\sum_{t=0}^{\infty} \delta^t u(\hat{y}_t)$ is finite then $e = (\mathbf{\rho}, \bar{y}, \delta, u)$ is admissible.

IV.1. *More Definitions and Notations*

We continue to denote the set of all programs *feasible* in e by $Z^*(e)$. An *evaluation criterion* $\mathcal{O}$ is a correspondence from E into $S^+ \times S^+ \times S^+$ such that, for all admissible e, $\mathcal{O}(e)$ is a non-empty subset of $Z^*(e)$. An evaluation criterion $\mathcal{O}^*$ is *optimal* over E if each element of $\mathcal{O}^*(e)$ is optimal (according to (4.1)) for *every* e in E. By using (A.4.1) it is easy to verify that if $Z^*(e)$ is convex for all e in E then $\mathcal{O}^*$ is single-valued (i.e., $\mathcal{O}^*(e)$ consists of a single program $(\mathbf{x}^*, \mathbf{y}^*, \mathbf{c}^*)$ in $Z^*(e)$) and non-wasteful. In what follows, we *assume that* this is indeed the case, i.e., that $\mathcal{O}^*(e)$ *has a single element for every e in E.* We use the notation $[\mathcal{O}^*(e)]_t$ to denote the unique triple (x_t^*, y_t^*, c_t^*) such that $(\mathbf{x}^*, \mathbf{y}^*, \mathbf{c}^*) = \mathcal{O}^*(e)$.

We define $\mathcal{O}^*$ to be *sensitive*[20] on E to a change in the technology in period $T(\geq 2)$ if there are in E two admissible environments $e = (\mathbf{f}, \bar{y}, \delta, u)$ and $\bar{e} = (\bar{\mathbf{f}}, \bar{y}, \delta, u)$ such that (i) $\mathbf{f}^{(T-1)} = \bar{\mathbf{f}}^{(T-1)}$, (ii) $f_T \neq \bar{f}_T$, and (iii) $[\mathcal{O}^*(e)]_0 \neq [\mathcal{O}^*(\bar{e})]_0$. It should be stressed that the envrionments e and $\bar{e}$ have the *same* initial stock $\bar{y}$, the same discount factor δ, and the same partial history of the technology up to $T-1$, i.e., $f_{t+1} = \bar{f}_{t+1}$ for $t = 0, \ldots, T-2$. Thus, roughly speaking, such sensitivity means that optimal consumption and input in period zero (from the same initial stock) vary due to a shift in the technology in period T (from f to $\bar{f}$), even though there is no change in the technological possibilities in all the periods up to $T-1$.[21]

Following our earlier terminology, we say that an optimal evaluation criterion $\mathcal{O}^*$ defined on a class of environments E is *realized* by a sequence (Ψ_t), $t = 0, 1, \ldots$, of (temporally) decentralized verification rules if (i) there is a sequence (T_t) of finite positive integers and (ii) there is a sequence (Ψ_t) of functions such that for any $\mathcal{O}^*$-admissible environment $e = (\mathbf{f}, \bar{y}, \delta, u)$ in E, we have $\Psi_t(\mathbf{f}^{(T)_t}, \bar{y}, \delta, u) = [\mathcal{O}^*(e)]_t$ for all $t \geq 0$. (If there exists such a sequence, we say that the criterion is (temporally) *decentralized* on E.)

Our next result asserts that if an optimal evaluation criterion is sensitive to a change in the technology in period T, one cannot decentralize it with a sequence (Ψ_t) such that Ψ_0 depends *only* on $(f_1, \ldots, f_{T_0+1})$ where $T_0 + 1 < T$.

R.4.1. *Suppose that a single-valued optimal evaluation criterion $\mathcal{O}^*$ is sensitive to a change in the technology in a particular period $T \geq 2$. Then $\mathcal{O}^*$*

[20] From now on unless specified to the contrary, "sensitive" means "sensitive in E."

[21] And there are no restrictions on f_τ versus $\bar{f}_\tau$ for $\tau > T$.

cannot be realized by any decentralized sequence (Ψ_t), $t=0, 1, \ldots$ *such that* $T_0+1<T$.

Proof. See Appendix (A.2).

Remark. If $T_0=0$, this result implies that if $\mathcal{O}^*$ is sensitive to a technology change even only in a single period $T\geqslant 2$, then $\mathcal{O}^*$ is not decentralizable. But it leaves open the possibility that $\mathcal{O}^*$ is decentralizable when $T_0+1\geqslant T$. With assumptions stronger than those in R.4.1, this possibility is ruled out by T.4 below.

T.4. *Suppose a single-valued optimal evaluation criterion* $\mathcal{O}^*$ *is sensitive to technology changes in some infinite sequence of periods. Then* $\mathcal{O}^*$ *is not decentralizable.*

Proof. The conclusion follows directly from R.4.1.

In view of T.4, it is natural to ask whether one can construct an example of an optimal evaluation criterion $\mathcal{O}^*$ that is sensitive to changes in the technology in an infinite sequence of periods. If the answer to this question is yes, the second question is whether such examples are "typical" or "exceptional." Example 4.1 provides a precise affirmative answer to the first question with sensitivity in *all* periods $T\geqslant 2$. The second question turns out to be quite subtle and is discussed in Section V.

EXAMPLE 4.1. Consider an infinite-horizon linear production model. Let $u(c)=c^\alpha/\alpha$ where $0<\alpha<1$. Clearly, u satisfies (A.4.1). Let E^* denote the class of environments satisfying the preceding conditions and admissible for (4.1). E^* is non-empty; e.g., E^* contains $(\boldsymbol{\rho}, \bar{y}, \delta, u)$ with $\boldsymbol{\rho}=(\rho_{t+1})$ where $\rho_{t+1}=1$ for all $t\in N$, $\bar{y}=1$, $\delta\in(0, 1)$. Choose any e in E^* where $e=(\boldsymbol{\rho}, \bar{y}, \delta, u)$ and $\boldsymbol{\rho}\in\mathring{S}$.

Consider the following optimiation problem:

$$\text{“given } y_0=\bar{y}>0, \text{ maximize } \sum_{t=0}^{\infty} \delta^t u(c_t) \text{ subject to}$$

$$\begin{aligned} &c_t+x_t\leqslant y_t, \qquad 0\leqslant y_{t+1}\leqslant \rho_{t+1}x_t \\ &c_t\geqslant 0, \quad x_t\geqslant 0 \qquad \text{for all } t\in N.\text{”} \end{aligned} \tag{P.1}$$

One can prove that there is a unique solution to (P.1) [and hence (P.1) defines on E^* a single-valued optimal evaluation criterion $\mathcal{O}^*$]. One can also prove that if $e=(\boldsymbol{\rho}, \bar{y}, \delta, u)$, $\mathcal{O}^*(e)\equiv(\mathbf{x}^*, \mathbf{c}^*)$ is interior (i.e., $c_t^*>0$ and $x_t^*>0$ for all $t\in N$) and satisfies the Ramsey–Euler condition:

$$u'(c_t^*)=\delta\rho_{t+1}u'(c_{t+1}^*) \qquad \text{for} \quad t\in N. \tag{4.4}$$

Also, since the optimal program is intertemporally efficient, we have (by (3.9)):

$$\sum_{t=0}^{\infty} c_t^*/\sigma_t = \bar{y}. \tag{4.5}$$

Consider another environment $\bar{e} \equiv (\bar{\boldsymbol{\rho}}, \bar{y}, \delta)$ such that $\bar{\boldsymbol{\rho}} \in \mathring{S}$ satisfies

$$\bar{\rho}_2 \neq \rho_2, \quad \bar{\rho}_t = \rho_t \qquad \text{for all } t \neq 2. \tag{4.6}$$

In other words, the output–capital ratio in period 2 is changed from ρ_2 to $\bar{\rho}_2$, keeping all *other* ρ_t, $\bar{y}$, and δ the same as in e. Since e is admissible so is $\bar{e}$. Write $\bar{\sigma}_0 \equiv 1$, $\bar{\sigma}_t \equiv \prod_{i=1}^{t} \bar{\rho}_i$. Then $\sigma_0 = \bar{\sigma}_0 = 1$; $\sigma_1 = \rho_1 = \bar{\rho}_1 = \bar{\sigma}_1$; and for all $t \geq 2$, $\bar{\sigma}_t = \sigma_t(\bar{\rho}_2/\rho_2)$. We now study the following optimisation problem:

$$\begin{aligned} &\text{“given } y_0 = \bar{y} > 0, \text{ maximize } \sum_{t=0}^{\infty} \delta^t u(c_t) \\ &\text{subject to} \quad c_t + x_t \leq y_t \\ &\qquad 0 \leq y_{t+1} \leq \bar{\rho}_{t+1} x_t; \\ &\qquad c_t \geq 0, \quad x_t \geq 0 \qquad \text{for} \quad t \in N.\text{”} \end{aligned} \tag{P.2}$$

Again, for any $\bar{y} > 0$, the unique solution to (P.2) (the optimal program in $\bar{e}$) $(\bar{\mathbf{x}}^*, \bar{\mathbf{c}}^*)$ is interior (i.e., $\bar{x}_t^* > 0$, $\bar{c}_t^* > 0$ for $t \in N$) and satisfies the Ramsey–Euler condition:

$$u'(\bar{c}_t^*) = \delta \bar{\rho}_{t+1} u'(\bar{c}_{t+1}^*) \qquad \text{for all } t \in N. \tag{4.7}$$

Since optimality implies intertemporal efficiency, one also has (by (3.9))

$$\sum_{t=0}^{\infty} \bar{c}_t^*/\bar{\sigma}_t = \bar{y}. \tag{4.8}$$

It is shown in Hurwicz and Majumdar [5] that the hypothesis $c_0^* = \bar{c}_0^*$ leads to a contradiction. This means that the criterion $\mathcal{O}^*$ *is* sensitive on E^* to technology change in the period $T = 2$. Indeed it can also be shown that $\mathcal{O}^*$ is also sensitive on E^* to technology change in any period $T > 2$. Hence, by T.4, the criterion $\mathcal{O}^*$ is not decentralizable on E^*.

V. Sensitivity of the Optimal Resource Allocation Mechanism and Decentralizability

V.1. *The Linear Model of Production*

Consider the class $E_{\mathscr{L},u}$ of environments $e = (\mathbf{f}, y, \delta, u)$ where u is a fixed felicity function satisfying (A.4.1), $y > 0$, $0 < \delta < 1$, and $\mathbf{f} \in \mathscr{L}^{(\infty)}$. Our main result (T.5) in this section is that the optimal evaluation criterion $\mathcal{O}^*$ is decentralized over $E_{\mathcal{L},u}$ if and only if the felicity function u is logarithmic (i.e., if $u(c)=k+\log c$, k a constant).

To begin with we *fix* a finite positive integer $T\geqslant 1$, and a discount factor δ in (0, 1). Consider the following[22] optimization problem (P.3):

$$\begin{aligned}&\text{“given } \bar{y}>0, \text{ maximize } \sum_{t=0}^{T} \delta^t u(c_t)\\ &\text{subject to } \sum_{t=0}^{T} c_t/\sigma_t \leqslant \bar{y};\quad c_t\geqslant 0 \qquad \text{for} \quad t=0, ..., T.\text{”}\end{aligned} \tag{P.3}$$

Choose a particular $\rho^*=(\rho_1^*, ..., \rho_T^*)\in P^T$. Let $c^{*(T)}=(c_0^*, ..., c_T^*)$ be the unique solution to (P.3). One can prove that $c_t^*>0$ for all t. Since $u'(c)>0$ for c in P, optimality implies intertemporal efficiency, and this leads to:

$$\sum_{t=0}^{T} c_t^*/\sigma_t^* = \bar{y}. \tag{4.9}$$

The first order conditions lead to the following system:

$$\delta^t u'(c_t^*) - u'(c_0^*)/\sigma_t^* = 0, \qquad t=1, ..., T; \tag{4.10}$$

$$\sum_{t=0}^{T} c_t^*/\sigma_t^* - \bar{y} = 0. \tag{4.11}$$

Applying the Implicit Function Theorem (see [5] for details) we differentiate (4.10) with respect to ρ_T and get the following (we write $\dot{c}_t^* \equiv \partial c_t^*/\partial \rho_T$):

$$\delta^t u''(c_t^*)\,\dot{c}_t^* - \frac{u''(c_0^*)\,\dot{c}_0^*}{\sigma_t^*} = 0, \qquad t=1, ..., T-1; \tag{4.12}$$

$$\delta^T u''(c_T^*)\,\dot{c}_T^* - \frac{(u''(c_0^*)\,\dot{c}_0^*)\,\sigma_T^* - u'(c_0^*)\,\sigma_{T-1}}{\sigma_T^{*2}} = 0. \tag{4.13}$$

[22] One can show that the problem (P.3) is equivalent to the following problem (P.3′) in which constraints appear in the same way as in our earlier definition of feasibility (see (3.2)):

$$\begin{aligned}&\text{“maximize } \sum_{t=0}^{T} \delta^t u(c_t)\\ &\text{subject to } y_0=\bar{y}\\ &\qquad c_t+x_t=y_t, \qquad t=0, ..., T;\\ &\qquad 0\leqslant y_{t+1}\leqslant \rho_{t+1}x_t, \qquad t=0, ..., T-1;\\ &\qquad c_t\geqslant 0,\quad x_t\geqslant 0, \qquad t=0, ..., T.\text{”}\end{aligned} \tag{P.3′}$$

Differentiating (4.11) with respect to ρ_T we get

$$\sum_{t=0}^{T-1} \dot{c}_t^*/\sigma_t^* + \frac{\dot{c}_T^*\sigma_T^* - c_T^*\sigma_{T-1}^*}{\sigma_T^{*2}} = 0. \tag{4.14}$$

If the optimal decision in period zero is not sensitive to changes in ρ_T, one has $\dot{c}_0^* = 0$. From (4.12), this leads to

$$\dot{c}_t^* = 0 \qquad \text{for} \quad t = 0, ..., T-1. \tag{4.15}$$

If $\dot{c}_0^* = 0$, we get from (4.13)

$$\delta^T u''(c_T^*)\, \dot{c}_T^* + \frac{u'(c_0^*)\, \sigma_{T-1}^*}{\sigma_T^{*2}} = 0. \tag{4.16}$$

Using (4.10) and (4.14) through (4.16) we get

$$\frac{u''(c_T^*)\, c_T^* \sigma_{T-1}}{\sigma_T^*} + \frac{u'(c_T^*)\, \sigma_{T-1}^*}{\sigma_T^*} = 0 \tag{4.17}$$

or

$$u''(c_T^*)\, c_T^* + u'(c_T^*) = 0. \tag{4.18}$$

Relation (4.22) can be derived on an open neighborhood of c^*. *Thus, if* $\dot{c}_0^* = 0$, *we must have* $u(c) = k + \log c$. On the other hand, if we start from $u(c) = k + \log c$, we can compute the optimal decisions, $c_t^* = \sigma_t \delta^t y/\bar{s}$ where $\bar{s} = \sum_{t=0}^{T} \delta^t$. Hence $\partial c_t^*/\partial \rho_{t'} = 0$ for $t' \geqslant t+1$. *Thus, insensitivity of the decisions in period zero with respect to changes in* ρ_T *obtains if and only if* $u(c) = k + \log c$.

The arguments mentioned above apply to *any* finite T. Suppose that the optimal evaluation criterion $\mathcal{O}^*$ can be realized by functions (Ψ_t), where Ψ_0 selects $[\mathcal{O}^*(\boldsymbol{\rho}, y, \delta)]_0 \equiv (x_0^*, y_0^*, c_0^*)$ as a function *only* of $(\boldsymbol{\rho}^{(T_0)}, y, \delta)$. Choose a particular $\boldsymbol{\rho}^* \in \mathring{S}$. Consider the environments $e(r) \equiv ((\rho_1^*, ..., \rho_{T_0}^*, r), y, \delta)$ where $r > 0$. Clearly, $\Psi_0(e(r))$ must be independent of variations in $r > 0$; i.e., $[\mathcal{O}^*(e(r))]_0$ must be independent of variations in r. By using the arguments spelled out above, we conclude that this can happen if and only if $u(c) = k + \log c$. To summarize, we have proved the following:

T.5.1. *Let* $\mathcal{O}^*$ *be the optimal evaluation criterion, defined by* (4.1) *over* $E_{\mathcal{L},u}$. *Then* $\mathcal{O}^*$ *is insensitive on* $E_{\mathcal{L},u}$ *to a change in the technology for every period* $T \geqslant 2$ *if and only if* $u(c) = k + \log c$, *where* k *is any constant.*

V.2. *Environments Containing Strictly Concave Gross Output Functions*

V.2.0. *Introduction*

We now turn to the question of sensitivity of the optimal evaluation criterion $\mathcal{O}^*$ when we allow the set J of all possible technologies in any period to contain neoclassical (hence, strictly concave) gross output functions.

Formally, let J contain at least one pair of (strictly concave) functions $\bar{f}$ and $\bar{F}$ in $\mathcal{F}$ satisfying (A.3.1)–(A.3.4) such that

(A.3.5) *For some* $b>0$, $\bar{f}(b)=\bar{F}(b)$ *and* $\bar{f}'(b)\neq\bar{F}'(b)$.

In other words, we require that at some input level b, the outputs resulting from technologies $\bar{f}$ and $\bar{F}$ be the same, but the marginal productivities differ. Note that if J contains any function $\bar{f}$ satisfying (A.3.1)–(A.3.4) and also contains $\bar{F}(x)=[\bar{f}(x)]^\alpha$ where $0<\alpha<1$, then (A.3.5) is satisfied.

Recall that $f^{(\infty)}$ denotes the stationary sequence $f^{(\infty)}=(f, f, \ldots)$. For any positive (finite) integer τ, we use the notation $\langle\bar{f}^{(\tau)}, \bar{F}^{(\infty)}\rangle$ to denote the sequence $\mathbf{f}=(f_{t+1})$ such that $f_{t+1}=\bar{f}$ for $t=0, \ldots, \tau-1$, and $f_{t+1}=\bar{F}$ for all $t\geqslant\tau$. In other words, $\langle\bar{f}^{(\tau)}, \bar{F}^{(\infty)}\rangle$ represents a sequence such that the gross output function changes from $\bar{f}$ to $\bar{F}$ in the period $\tau+1$. Define the set $J(\bar{f}, \bar{F})$ as

$$J(\bar{f}, \bar{F})=\{\mathbf{f}\colon \mathbf{f}=\langle\bar{f}^{(\tau)}, \bar{F}^{(\infty)}\rangle \text{ for some positive (finite) integer } \tau,$$
$$\text{or } \mathbf{f}=\bar{f}^{(\infty)}, \text{ or } \mathbf{f}=\bar{F}^{(\infty)}\}.$$

Let E be a class of environments $e=(\mathbf{f}, y, \delta, u)$ in which $y\in P$, $\mathbf{f}$ ranges over a subset $J'^{(\infty)}$ of $J^{(\infty)}$ containing $J(\bar{f}, \bar{F})$ where the pair $(\bar{f}, \bar{F})$ satisfies (A.3.1)–(A.3.5), while δ and u are fixed, with $\delta\in(0, 1)$ and u satisfying (A.4.1).

Consider the following dynamic optimization problem (P.4):

$$\begin{aligned}
&\text{“maximize} && \sum_{t=0}^{\infty}\delta^t u(c_t) && \\
&\text{subject to} && c_0+x_0=\bar{y}; && \\
& && c_t+x_t=f_t(x_{t-1}) && \text{for } t\geqslant 1; \\
& && c_t\geqslant 0,\quad x_t\geqslant 0 && \text{for } t\in N; \qquad \text{(P.4)}\\
&\text{given} && \bar{y}>0,\quad 0<\delta<1, && f_t\in J.\text{”}
\end{aligned}$$

As the proof of T.5.2, the central result of this Section, is long and technical, it is perhaps useful to describe it informally. Suppose that any integer $\tau\geqslant 1$ is given. T.5.2 asserts the existence of an initial stock $y_0>0$

and two admissible environments e and e' in E such that $[\mathcal{O}^*(e)]_0 \neq [\mathcal{O}^*(e')]_0$, i.e., such that optimal consumption and input in period zero differ between e and e'. (The environments e and e' are constructed as $e \equiv (\hat{f}^{(\infty)}, y_0, \delta, u)$ and $e' \equiv (\langle \hat{f}^{(\tau)}, \bar{F}^{(\infty)} \rangle, y_0, \delta, u)$, where $\hat{f}$ and $\bar{F}$ are specified in (A.3.5).) Thus the optimal evaluation criterion $\mathcal{O}^*$ on E is sensitive to a change in the technology in *every* period $\tau \geqslant 1$. In view of T.4.1, this implies that $\mathcal{O}^*$ cannot be decentralized on E.

The proof of T.5.2 is contained in Section V.2.3. It is based on auxiliary results given in Sections V.2.1 and V.2.2.

V.2.1. *Dynamic Optimization in a Stationary Environment*

We recall some properties of the optimal program in the environment $e \equiv (f^{(\infty)}, y, \delta, u)$ where δ is a fixed number in (0, 1), f is *any* neoclassical gross output function (satisfying (A.3.1)–(A.3.4)) assumed to belong to J, and y is *any* positive initial stock. Consider the following optimization problem:

$$\begin{array}{lllr} \text{"maximize} & \sum_{t=0}^{\infty} \delta^t u(c_t) & & \\ \text{subject to} & c_0 + x_0 = y & & \\ & c_t + x_t = f(x_{t-1}) & \text{for } t \geqslant 1 & \text{(P.5)} \\ & c_t \geqslant 0, \quad x_t \geqslant 0 & \text{for } t \in N. & \\ \text{given} & y > 0, \quad 0 < \delta < 1." & & \end{array}$$

Note that this problem (P.5) is the special case of our earilier problem (P.4) where $f_t \equiv f$ for all $t \geqslant 1$. One can verify that for *any* $y > 0$ there is a unique optimal program $(\mathbf{x}^*, \mathbf{y}^*, \mathbf{c}^*)_y$ in the environment $e \equiv (f^{(\infty)}, y, \delta, u)$; i.e., the problem (P.5) has a unique solution. The following result is well known from the dynamic programming literature:

R.5.1. *Given the function f and $\delta \in (0, 1)$ there is a unique function $h\colon P \to P$ with the following properties:*

(i) $0 < h(y) < y$.

(ii) *For any $y > 0$, define $(\mathbf{x}^*, \mathbf{y}^*, \mathbf{c}^*)$ as*

$$c_0^* = h(y), \qquad x_0^* = y - h(y);$$

$$c_t^* = h(y_t^*), \qquad x_t^* = y_t^* - h(y_t^*) \qquad \textit{for all } t \geqslant 1;$$

$$y_t^* = f(x_{t-1}^*) \qquad \textit{for all } t \geqslant 1.$$

Then $(\mathbf{x}^*, \mathbf{y}^*, \mathbf{c}^*)$ is the unique optimal program in $e = (f^{(\infty)}, y, \delta, u)$.

(iii) *h is strictly increasing; i.e.,* $y' > y''$ *implies* $h(y') > h(y'')$.

(iv) *h is continuous on P.*

(v) $h(y)$ *tends to zero as y tends to zero.*

We refer to h as the *optimal consumption policy function.* Define the function $i\colon P \to P$ as

$$i(y) \equiv y - h(y). \tag{5.1}$$

We refer to $i\colon P \to P$ as the *optimal investment policy function.* One can also derive the Ramsey–Euler condition (see, e.g., Mirman and Zilcha [11]):

$$u'(h(y)) = \delta f'[y - h(y)]\, u'[h(f(y - h(y)))]. \tag{5.2}$$

Equivalently, using (5.1), we can also write (5.2) as

$$u'(h(y)) = \delta f'(i(y))\, u'[h(f(i(y)))]. \tag{5.3}$$

The following property of h corresponding to the chosen (f, δ) is noted:

R.5.2. $h(y)$ *goes to infinity as y goes to infinity.*

Next, we want to summarize some properties of the optimal investment policy function i.

R.5.3. (i) $i(y)$ *goes to infinity as y tends to infinity.*

(ii) *i is strictly increasing; i.e.,* $y' > y''$ *implies* $i(y') > i(y'')$.

(iii) *i is continuous on P.*

(iv) $i(y)$ *tends to zero as y tends to zero.*

An immediate consequence of the results stated above is the following:

R.5.4. *Given any* $\hat{\beta} > 0$, *there is a unique initial stock* $y^* > 0$ *such that*

$$h(y^*) = y^* - \hat{\beta} \text{ [equivalently, } i(y^*) = \hat{\beta}].$$

Remark. It should perhaps be stressed that the functions h and i depend on the gross output function f and the discount factor δ in (0, 1) that have been chosen. Also, the number y^* in R.5.4 depends on $\hat{\beta}$[23].

V.2.2. *Comparison of Optimal Policy Functions*

The results obtained in the last Section V.2.1 can now be applied to two distinct stationary environments: $(\bar{f}^{(\infty)}, y, \delta, u)$ and $(\bar{F}^{(\infty)}, y, \delta, u)$ where $\bar{f}$

[23] For proofs of R.5.3 and R.5.4 the interested reader may turn to [5].

and $\bar{F}$ are the strictly concave gross output functions specified in (A.3.5). Let $\bar{h}$ (resp. $\bar{i}$) be the optimal consumption (resp. investment) policy function that one derives from (P.5) with $f=\bar{f}$. Similarly, let $\bar{H}$ be the optimal consumption policy function derived from (P.5) with $f=\bar{F}$. Both $\bar{h}$ and $\bar{H}$ have all the properties listed in R.5.1 and R.5.2. We now prove the intuitive result: the functions $\bar{h}$ and $\bar{H}$ cannot be identical. In other words, there is some initial stock $\bar{y}>0$ such that the optimal consumption $\bar{h}(\bar{y})$ in the environment $(\bar{f}^{(\infty)}, \bar{y}, \delta, u)$ is different from the optimal consumption $\bar{H}(\bar{y})$ in the environment $(\bar{F}^{(\infty)}, \bar{y}, \delta, u)$.[24]

R.5.5. *Let $\bar{h}$ and $\bar{H}$ be the optimal consumption policy functions obtained from* (P.5) *by setting $f=\bar{f}$ and $f=\bar{F}$, respectively. There is some $\bar{y}>0$ such that*

$$\bar{h}(\bar{y}) \neq \bar{H}(\bar{y}). \tag{5.4}$$

Proof. See Appendix (A.3).

V.2.3. *The Optimal Evaluation Criterion is Sensitive in all Periods*

Let $\mathcal{O}^*$ be the optimal resource allocation mechanism and consider the stationary environment $e=(\bar{f}^{(\infty)}, y, \delta, u)$ where $y>0$. The unique optimal program $(\mathbf{x}^*, \mathbf{y}^*, \mathbf{c}^*)$ from y is selected by $\mathcal{O}^*$; i.e.,

$$\mathcal{O}^*(e) \equiv (\mathbf{x}^*, \mathbf{y}^*, \mathbf{c}^*) = \{(x_t^*)_{t=0}^{\infty}, (y_t^*)_{t=0}^{\infty}, (c_t^*)_{t=0}^{\infty}\}.$$

One notes that for every $T \geq 1$, the program $\{(x_t^*)_{t=T}^{\infty}, (y_t^*)_{t=T}^{\infty}, (c_t^*)_{t=T}^{\infty}\}$ is the unique optimal program in the environment $(\bar{f}^{(\infty)}, y_T^*, \delta, u)$. More explicitly, consider the following optimization problem:

$$\begin{aligned} &\text{“maximize } \sum_{t=0}^{\infty} \delta^t u(c_t) \\ &\text{subject to } \quad c_0 + x_0 = y_T^* \\ &\qquad c_t + x_t = \bar{f}(x_{t-1}) \qquad \text{for } \quad t \geq 1; \\ &\qquad c_t \geq 0, \quad x_t \geq 0 \qquad \text{for } \quad t \in N.\text{”} \end{aligned} \tag{P.6}$$

We emphasize that the initial stock in problem (P.6) is equal to $y_T^*>0$. The unique solution to problem (P.6) denoted by $(\bar{\mathbf{x}}^*, \bar{\mathbf{y}}^*, \bar{\mathbf{c}}^*)$ is given by

$$\bar{x}_t^* = x_{T+t}^*, \quad \bar{y}_t^* = y_{T+t}^*, \quad \bar{c}_t^* = c_{T+t}^* \qquad \text{for all } t \geq 0.$$

Let τ be any positive integer. Consider the environment

[24] The idea behind the proof of R.5.5 is due to Tapan Mitra.

$e' = (\langle \bar{f}^{(\tau)}, \bar{F}^{(\infty)} \rangle, y, \delta, u)$. In e' the gross output function is $\bar{f}$ in periods 1 through τ, but it is $\bar{F}$ from the period $\tau+1$ onward. Again, there is a unique optimal program in e' which we denote by $\langle \mathbf{x}', \mathbf{y}', \mathbf{c}' \rangle$. One notes that $\{(x_t')_{t=\tau+1}^{\infty}, (y_t')_{t=\tau+1}^{\infty}, (c_t')_{t=\tau+1}^{\infty}\}$ is the unique optimal program in the environment $e'' = (\bar{F}^{(\infty)}, y_{\tau+1}', \delta)$. Our main result can now be stated and proved:

T.5.2. *Let τ be any positive (finite) integer. There is some initial stock $y_0(\tau) > 0$ such that for the admissible environments $e(\tau) = (\bar{f}^{(\infty)}, y_0(\tau), \delta, u)$ and $e'(\tau) = (\langle \bar{f}^{(\tau)}, \bar{F}^{(\infty)} \rangle, y_0(\tau), \delta, u)$ in E one has*

$$[\mathcal{O}^*(e(\tau))]_0 \neq [\mathcal{O}^*(e'(\tau))]_0. \tag{5.5}$$

Remark. T.5.2 means that the criterion $\mathcal{O}^*$ is sensitive on E to technology change in every period. Therefore, by T.4, $\mathcal{O}^*$ is not decentralizable on E.

Proof. See Appendix (A.4).

VI. Strong Sensitivity

In this section we analyse the implications of a stronger notion of sensitivity.[25] We begin by specifying a new class $\bar{E}$ of a priori admissible environments. Define $\bar{J}(\bar{f}, \bar{F})$ as follows:

$$\bar{J}(\bar{f}, \bar{F}) = \{\mathbf{f}: \mathbf{f} = (\bar{f}^{(\infty)}) \text{ or } \mathbf{f} = \langle \underbrace{\bar{f}, \ldots, \bar{f}}_{\tau\text{-times}}, \bar{F}, \bar{f}^{(\infty)} \rangle \text{ for any finite positive integer } \tau \geq 1\}.$$

Let $\bar{E}$ be any set of environments $e = (\mathbf{f}, y, \delta, u)$ where $\mathbf{f}$ ranges over any subset of $J^{(\infty)}$ containing $\bar{J}(\bar{f}, \bar{F})$, $y \in P$, $\delta \in (0, 1)$ and u satisfies (A.4.1). Let $\mathcal{O}^*$ be the *optimal evaluation criterion over* $\bar{E}$. We say that $\mathcal{O}^*$ is *strongly sensitive* in $\bar{E}$ to a change in the technology in period $\tau \geq 2$ if there exists two admissible environments $e(\tau) = (\mathbf{f}, y(\tau), \delta, u)$ and $\hat{e}(\tau) = (\hat{\mathbf{f}}, y(\tau), \delta, u)$ such that (i′) $f_t = \hat{f}_t$ for all $t \neq \tau$, (ii) $f_\tau \neq \hat{f}_\tau$ and (iii) $[\mathcal{O}^*(e(\tau))]_0 \neq [\mathcal{O}^*(\hat{e}(\tau))]_0$.

Note the following:

(a) Both $e(\tau)$ and $\hat{e}(\tau)$ have the same initial stock and discount factor, and the same technology in all but one period (namely the τth period).

[25] The impossibility results reported in this section were derived in response to comments made by Professor H. Weinberger regarding the finiteness restrictions on the domain of the verifications rules (ψ_t) that were made in Section IV.

(b) Strong sensitivity *implies* the sensitivity concept introduced in Section IV.1 above, but is *not* implied by it.

One can prove the following (see Hurwicz and Majumdar [6] for details; the arguments are similar to those leading to T.5.2):

T.6.1. *The optimal evaluation criterion $\mathcal{O}^*$ is strongly sensitive on $\bar{E}$ to a change in the technology in all periods $t \geq 2$.*

In T.5.2 the initial phase of the optimal program is shown to vary as a result of a *permanent* change from one technology to another. In T.6.1, the variation is in response even to a temporary (one period) change in technology. The two phenomena may be of relevance in different applied situations, but—as it turns out—both result in the impossibility of designing intertemporally decentralized mechanisms guaranteeing optimality.

Let M be any proper subset (*finite or infinite*) of N containing 0 and not containing some integer $\tau \geq 2$. Given any sequence $\mathbf{f} = (f_{t+1})_{t \in N}$ of gross output functions, denote by $\mathbf{f}^{(M)} = (f_{t+1})_{t \in M}$ the restriction of the sequence to M. For any admissible environment $e = (\mathbf{f}, y, \delta, u)$ let ψ_0 be any verification rule defined on $e^{(M)} = (\mathbf{f}^{(M)}, y, \delta, u)$. In other words, ψ_0 is restricted to using information about the technologies in a proper (*finite or infinite*) subset M of N. From T.6.1 one can readily obtain:

T.6.2. *Non-decentralizability: Let $\mathcal{O}^*$ be the optimal evaluation criterion over $\bar{E}$. Given any proper subset M (finite or infinite) of N, containing 0 and not containing some integer $\tau \geq 2$, $\mathcal{O}^*$ cannot be realized by any sequence (ψ_t) of decentralized verification rules where ψ_0 is defined on $e^{(M)}$ for all $e \in \bar{E}$.*

Appendix

(A.1) *Proof of* T.3. Suppose, to the contrary, that there exist a sequence (T_t) of finite positive integers and a sequence $\Psi_t : (\mathbf{f}^{(T_t)}, \bar{y}) \to \bar{P}^3$ of correspondences such that $(\mathbf{x}, \mathbf{y}, \mathbf{c})_{\bar{y}}$ is efficient if and only if (x_t, y_t, c_t) is in $\Psi_t(f^{(T_t)}, \bar{y})$ for all t. We show that this supposition leads to a contradiction.

Consider the class of (single period consumption) efficient programs $(\mathbf{x}^T, \mathbf{y}^T, \mathbf{c}^T)_{\bar{y}}$, $T = 1, 2, \ldots$, as defined in (3.5). Since $(\mathbf{x}^1, \mathbf{y}^1, \mathbf{c}^1)_{\bar{y}}$ is efficient, our hypothesis requires that (x_0^1, y_0^1, c_0^1) be in $\Psi_0(\mathbf{f}^{(T_0)}, \bar{y})$. From (3.5) we observe that $x_0^1 = y_0^1 = \bar{y}$; $c_0^1 = 0$; hence, $(\bar{y}, \bar{y}, 0)$ is in $\Psi_0(\mathbf{f}^{(T_0)}, \bar{y})$. Next, consider the efficient program $(\mathbf{x}^2, \mathbf{y}^2, \mathbf{c}^2)_{\bar{y}}$. Again, according to our hypothesis, (x_1^2, y_1^2, c_1^2) is in $\psi_1(f^{(T_1)}, \bar{y})$. From (3.5) we observe that $x_1^2 = y_1^2 = f_1(\bar{y})$; $c_1^2 = 0$. Hence $(f_1(\bar{y}), f_1(\bar{y}), 0)$ is in $\psi_1(\mathbf{f}^{(T_1)}, \bar{y})$. Similarly, $(f_2 \cdot f_1(\bar{y}),$

$f_2 \cdot f_1(\bar{y}), 0)$ is in $\psi_2(\mathbf{f}^{(T_2)}, \bar{y})$. Proceeding this way, we conclude that $(f_t \cdot f_{t-1} \cdot \cdots \cdot f_1(\bar{y}), f_t \cdot f_{t-1} \cdot \cdots \cdot f(\bar{y}), 0)$ is in $\Psi_t(\mathbf{f}^{(T_t)}, \bar{y})$ for *all* t. Going back to the definition (3.3) of the program $(\hat{x}, \hat{y}, \hat{c})_{\bar{y}}$ of pure accumulation, we observe that, since $(\hat{x}_t, \hat{y}_t, \hat{c}_t) \equiv (f_t \cdot f_{t-1} \cdot \cdots \cdot f_1(\bar{y}), f_t \cdot f_{t-1} \cdot \cdots \cdot f_1(\bar{y}), 0)$ is in $\Psi_t(\mathbf{f}^{(T_t)}, \bar{y})$ for all t, our hypothesis implies that the (inefficient) program of pure accumulation $(\hat{\mathbf{x}}, \hat{\mathbf{y}}, \hat{\mathbf{c}})_{\bar{y}}$ must be in $Z^{**}(\mathbf{f}, y)$. This contradiction establishes the result.

(A.2) *Proof of* R.4.1. Since $\mathcal{O}^*$ is sensitive to technology change in period $T > T_0 + 1$ there are in E two admissible environments $\bar{e} = (\bar{\mathbf{f}}, \bar{y}, \delta, u)$ and $\bar{\bar{e}} = (\bar{\bar{\mathbf{f}}}, \bar{y}, \delta, u)$ such that $\bar{\mathbf{f}} = (\bar{f}_{t+1})$ and $\bar{\bar{\mathbf{f}}} = (\bar{\bar{f}}_{t+1})$ have the properties: $\bar{f}_{t+1} = \bar{\bar{f}}_{t+1}$ for $t = 0, \ldots, T-2$, $\bar{f}_T \neq \bar{\bar{f}}_T$ where $T > T_0 + 1$ and $[\mathcal{O}^*(\bar{e})]_0 \neq [\mathcal{O}^*(\bar{\bar{e}})]_0$. If $\mathcal{O}^*$ is realized by (Ψ_t) such that $\Psi_0(\mathbf{f}^{(T_0)}, \bar{y}, \delta, u) = [\mathcal{O}^*(e)]_0$ for all admissible e, we have $[\mathcal{O}^*(\bar{e})]_0 = \Psi_0(\bar{f}_1, \ldots, \bar{f}_{T_0+1}, \bar{y}, \delta, u) = \Psi_0(\bar{\bar{f}}_1, \ldots, \bar{\bar{f}}_{T_0+1}, \bar{y}, \delta, u) = [\mathcal{O}^*(\bar{\bar{e}})]_0$, a contradiction.

(A.3) *Proof of* R.5.5. Assume, to the contrary, that $\bar{h}(y) \equiv \bar{H}(y)$ for all $y > 0$. We show that this leads to a contradicion. From the Ramsey–Euler conditions (5.2) for the problems (P.5) with $f = \bar{f}$ and $f = \bar{F}$ we get

$$u'[\bar{h}(y)] = \delta \bar{f}'[y - \bar{h}(y)]\, u'[\bar{h}\{\bar{f}(y - \bar{h}(y))\}] \tag{a.1}$$

and

$$u'[\bar{H}(y)] = \delta \bar{F}'[y - \bar{H}(y)]\, u'[\bar{H}\{\bar{F}(y - \bar{H}(y))\}]. \tag{a.2}$$

As $\bar{h}(y) \equiv \bar{H}(y)$ we can rewrite (a.2) as

$$u'[\bar{h}(y)] = \delta \bar{F}'[y - \bar{h}(y)]\, u'[\bar{h}\{\bar{F}(y - \bar{h}(y))\}]\,. \tag{a.3}$$

Using R.5.4, choose $\bar{y}^*$ to satisfy $\bar{h}(\bar{y}^*) = \bar{y}^* - b$ (where b is specified in (A.3.5)), and from (A.3.5) recall that

$$\bar{f}(\bar{y}^* - \bar{h}(\bar{y}^*)) = \bar{f}(b) = \bar{F}(b) = \bar{F}(\bar{y}^* - \bar{h}(\bar{y}^*)). \tag{a.4}$$

Also, again by (A.3.5), $\bar{F}'(b) \neq \bar{f}'(b)$.

From (a.1), at $y = \bar{y}^*$ one gets

$$u'[\bar{h}(\bar{y}^*)] = \delta \bar{f}'(b)\, u'[\bar{h}(\bar{f}(b))]. \tag{a.5}$$

From (a.3), at $y = \bar{y}^*$ one has

$$u'[\bar{h}(\bar{y}^*)] = \delta \bar{F}'(b)\, u'[\bar{h}(\bar{F}(b))]. \tag{a.6}$$

From (a.5) and (a.6) one gets $\bar{f}'(b) = \bar{F}'(b)$, contradicting (a.4). Thus, there must be some $\bar{y} > 0$ such that $\bar{h}(\bar{y}) \neq \bar{H}(\bar{y})$.

(A.4) *Proof of* T.5.2. We write out the proof in detail for the Case (A) where $\tau = 1$. The Case (B) where $\tau \geqslant 2$ is treated in a similar manner, and is sketched in the last paragraph of the proof.

(A) *The case* $\tau = 1$. Let $\bar{y} > 0$ be the initial stock whose existence is asserted in R.5.5 and consider

$$\bar{f}[\bar{\imath}(y)] = \bar{y} > 0, \qquad y > 0. \tag{a.7}$$

We recall that (i) $\bar{f}$ is continuous, strictly increasing; (ii) $\bar{f}(0) = 0$ and $\bar{f}(x)$ tends to infinity as x tends to infinity; (iii) $\bar{\imath}$ is continuous, strictly increasing; and (iv) $\bar{\imath}(y)$ tends to zero as y goes to zero, $\bar{\imath}(y)$ tends to infinity as y tends to infinity. Hence, we assert that there is a unique solution $y_0(1) > 0$ to (a.7) given by

$$y_0(1) = (\bar{\imath})^{-1}[(\bar{f})^{-1}(\bar{y})]. \tag{a.8}$$

Suppose that $(\mathbf{x}^*, \mathbf{y}^*, \mathbf{c}^*)$ is optimal in $e(1) \equiv (\bar{f}^{(\infty)}, y_0(1), \delta, u)$ and $(\mathbf{x}', \mathbf{y}', \mathbf{c}')$ is optimal in $e'(1) \equiv (\langle \bar{f}, \bar{F}^{(\infty)} \rangle, y_0(1), \delta, u)$. *We show that the hypothesis* $c_0^* = c_0'$, *leads to a contradiction.* If $c_0^* = c_0'$, one has

$$x_0^* = y_0(1) - c_0^* = y_0(1) - c_0' = x_0' \tag{a.9}$$

$$y_1^* = \bar{f}(x_0^*) = \bar{f}(x_0') = y_1' = \bar{y}. \tag{a.10}$$

Also,

$$u'(c_1^*) = u'(c_0^*)/\delta\bar{f}'(x_0^*) = u'(c_0')/\delta\bar{f}'(x_0') = u'(c_1') \tag{a.11}$$

From (a.11), we get

$$c_1^* = c_1'. \tag{a.12}$$

However, $\langle (x_t^*)_{t=1}^{\infty}, (y_t^*)_{t=1}^{\infty}, (c_t^*)_{t=1}^{\infty} \rangle$ and $\langle (x_t')_{t=1}^{\infty}, (y_t')_{t=1}^{\infty}, (c_t')_{t=1}^{\infty} \rangle$ are both optimal programs from the same initial $\bar{y}$ in the environments $(f^{(\infty)}, \bar{y}, \delta, u)$ and $(\bar{F}^{(\infty)}, \bar{y}, \delta, u)$, respectively. Hence according to (5.4)

$$c_1^* = \bar{h}(\bar{y}) \neq \bar{H}(\bar{y}) = c_1' \tag{a.13}$$

which contradicts (a.12). Hence, we get (5.5) for $\tau = 1$.

(B) *The case* $\tau \geqslant 2$. For any positive integer $\tau \geqslant 2$, write $(\bar{f} \cdot \bar{\imath})^{(\tau)}(y)$ to denote the composition of $(\bar{f} \cdot \bar{\imath})$ with itself τ times (e.g., $(f \cdot i)^{(2)}(y) \equiv \bar{f}(\bar{\imath}(\bar{f}(\bar{\imath}(y))))$, etc.). In order to complete the proof for any $\tau \geqslant 2$, one considers

$$(\bar{f} \cdot \bar{\imath})^{(\tau)}(y) = \bar{y}, \qquad y > 0, \tag{a.14}$$

where (as in (a.7) $\bar{y}$ is the initial stock whose existence is proved in R.5.5. By using the arguments leading to (a.9) one can also assert that (a.14) has a unique solution $y_0(\tau) > 0$. Next, consider the environments $e(\tau) = (\bar{f}^{(\infty)}, y_0(\tau), \delta, u)$ and $e'(\tau) = (\langle \bar{f}^{(\tau)}, \bar{F}^{(\infty)} \rangle, y_0(\tau), \delta, u)$. The method of proof spelled out above can be adapted to establish that $[\mathcal{O}^*(e(\tau))]_0 \neq [\mathcal{O}^*(e'(\tau))]_0$.

REFERENCES

1. P. DASGUPTA AND G. HEAL, "Economic Theory of Exhaustible Resources," Cambridge Univ. Press, London, 1979.
2. L. HURWICZ, Optimality and informational efficiency in resource allocation processes, *in* "Mathematical Methods in the Social Sciences" (K. Arrow, S. Kartin, and P. Suppes, Eds.), pp. 27–46, Stanford Univ. Press, Stanford, 1960.
3. L. HURWICZ, On informationally decentralized systems, *in* "Decision and Organization" (C. B. McGuire and R. Radner, Eds.), pp. 207–336, North-Holland, Amsterdam, 1972.
4. L. HURWICZ, On informational decentralization and efficiency in resource allocation mechanisms, *in* "Stud. Math. Econ., MAA Stud. Math.", (S. Reiter, Ed.), pp. 238–250, The Mathematical Association of America, Vol. 25, 1985.
5. L. HURWICZ AND M. MAJUMDAR, Optimal intertemporal allocation mechanisms and decentralization of decisions, Working Paper No. 369, Department of Economics, Cornell University, Ithaca, NY, 1985.
6. L. HURWICZ AND M. MAJUMDAR, On sensitivity and decentralization in infinite horizon optimization models, Working Paper No. 368, Deparment of Econmics, Cornell University, Ithaca, NY, 1985.
7. T. C. KOOPMANS, "Three Essays on the State of Economic Science," McGraw–Hill, New York, 1957.
8. M. MAJUMDAR, Efficient programs in infinite dimensional spaces: A complete characterization, *J. Econ. Theory* **7** (1974), 355–369.
9. M. MAJUMDAR, Some remarks on optimal growth with intertemporally dependent preferences in the neoclassical model, *Rev. Econ. Stud.* **XLII** (1975), 147–153.
10. E. MALINVAUD, Capital accumulation and efficient allocation of resources, *Econometrica* **21** (1953), 233–268.
11. L. MIRMAN AND I. ZILCHA, On optimal growth under uncertainty, *J. Econ. Theory* **11** (1975), 329–339.
12. T. MITRA AND M. MAJUMDAR, A note on the role of transversality condition in signaling capital over accumulation, *J. Econ. Theory* **13** (1976), 47–57.
13. K. MOUNT AND S. REITER, The informational size of message spaces, *J. Econ. Theory* **8** (1974), 161–192.
14. H. NIKAIDO, "Convex Structure and Economic Theory," Academic Press, New York, 1968.

3

On Characterizing Optimal Competitive Programs in Terms of Decentralizable Conditions*

WILLIAM A. BROCK

Department of Economics, University of Wisconsin, Madison, Wisconsin 53706

AND

MUKUL MAJUMDAR

Department of Economics, Cornell University, Ithaca, New York 14853

The paper considers a multisector model of intertemporal allocation with a primary factor of production and the overtaking criterion of optimality. The optimal program is characterized in terms of (i) period-by-period conditions on intertemporal profit and utility maximization relative to a system of competitive prices and (ii) non-positivity of appropriately computed values of differences of stocks from the golden rule stock: The last condition replaces the usual transversality condition of Malinvaud and throws new light on the possibility of dencentralization in an infinite-horizon economy. *Journal of Economic Literature* Classification Numbers: 020, 022, 027, 110, 111, 112, 113.

I. INTRODUCTION

This note is motivated by the recent work of Hurwicz and Majumdar [6] on designing informationally decentralized resource allocation mechanisms that are optimal in an infinite-horizon economy. In the well-known literature on the theory of intertemporal resource allocation (developing out of Ramsey [16] and Malinvaud [7]), optimality is

* This is a revised version of our earlier manuscript prepared (in November 1984) during Mukul Majumdar's stay at the University of Wisconsin, Madison. Research support from the National Science Foundation is gratefully acknowledged. Thanks are due to Professors L. Hurwicz, T. Mitra, S. Dasgupta and D. Ray for extremely helpful conversations, and to William Dean for a careful reading of the manuscript. Suggestions from the referees have improved the exposition.

characterized in terms of (i) period-by-period conditions on intertemporal profit and utility maximization relative to a system of competitive prices and (ii) a transversality condition on the sequence of values of inputs computed at these prices (the condition requires that in the "discounted" model this sequence must go to zero and that in the "undiscounted" model this sequence must remain uniformly bounded). It has been duly noted that it is difficult to contemplate an informationally decentralized mechanism (which exhibits initial dispersion of information and limited communication in the sense of Hurwicz [5])[1] that can ensure the fulfillment of the appropriate transversality condition by supplementing the competitive prices and maximization rules with messages that are finite dimensional or at least based only on a partial history of the environment in the dynamic model.

It turns out that when the technology is stationary, it is indeed possible to characterize optimality in terms of the usual competitive conditions (developed by Gale [4] and others) and yet another period-by-period condition that involves two other parameters. In Sections II and III we discuss the *undiscounted* case in detail. Assume that the time-invariant technology admits a unique "golden rule" equilibrium $((x^*, y^*, c^*), p^*)$ [see (3.3) and (3.4) below for the definition]. Our main results can be roughly summarized as follows: under appropriate strict convexity assumptions, a program $(\mathbf{x}, \mathbf{y}, \mathbf{c})$ from $y_0 \gg 0$ is optimal if and only if (i) it is competitive at prices $\mathbf{p} = (p_t)$ [see (2.8) and (2.9)] and (ii) $(y_t - y^*)(p_t - p^*) \leqslant 0$ for all $t \geqslant 0$. Note that a verification of condition (ii) requires the knowledge of y_t, p_t, *and* the golden rule stock y^* and the price system p^* supporting the golden rule.

In Section IV, some informal remarks on possible extensions of our characterization to other models are made. No attempt is made to provide a complete list of related models and results: the interested reader is referred to the extensive bibliographies in McKenzie [9] and Cass and Majumdar [1].

II. Programs, Price, and Optimality

II.a. *Preliminaries*

Let R^n be the n-dimensional Euclidean space; if $x = (x_i) \in R^n$, we write $x \geqslant 0$ (x is *non-negative*) if $x_i \geqslant 0$ for all $i = 1, 2, ..., n$; $x > 0$ (x is *semi-*

[1] Informational decentralization involves two basic restrictions: first, that there is initial dispersion of information with each economic unit processing only partial knowledge of the environment, and second, that there is limited communication in the sense that it is impossible through communication to completely centralize dispersed information so that some unit would have complete information about the environment.

positive) if $x \geqslant 0$ *and* $x \neq 0$; $x \gg 0$ (x is *strictly positive*) if $x_i > 0$ for all $i = 1, 2, ..., n$. $R^n_+ = [x \in R^n : x \geqslant 0]$ and $R^n_{++} = [x \in R^n : x \gg 0]$. For $x \in R^n$, $\|x\| = \sum_{i=1}^n |x_i|$. N is the set of all non-negative integers, i.e., $N = \{0, 1, 2, ...\}$. We denote by s the linear topological space of all real sequences of n-vectors (endowed with the topology of coordinatewise convergence).

II.b. *The Model*

We use the stock version of the multisector model of production spelled out in Nikaido [11, Chap. 4]. Only the relevant formal definitions and assumptions will be introduced. There are *n producible* goods and a single non-producible *primary factor of production* (called "labor"). Labor is used as an input in production, but does not enter into consumption. The supply of labor in period t, denoted by L_t, is given by

$$L_t = L_0 \lambda^t, \qquad L_0 > 0, \quad \lambda > 0; \quad t \in N. \tag{2.1}$$

An activity is a triplet $(L, X, Y) \in R_+ \times R^n_+ \times R^n_+$, where L is the quantity of labor input, X the vector of inputs of producible goods, and Y the vector of outputs of producible goods. Let $J' \subset R_+ \times R^n_+ \times R^n_+$ be the set of all technologically feasible activities. The following assumptions on J' are made:

(T′.1) *J' is a closed convex cone containing* $(0, 0, 0)$ [constant returns to scale]; *moreover*, "$(0, 0, y) \in J'$" *implies* "$y = 0$" [impossibility of free production].

(T′.2) "$(L, X, Y) \in J'$, $L' \geqslant L$, $X' \geqslant X$, $0 \leqslant Y' \leqslant Y$" *implies* "$(L', X', Y') \in J'$" [free disposal].

(T′.3) *There exists* $(\hat{L}, \hat{X}, \hat{Y}) \in J'$ *such that* $\hat{Y} \gg \lambda \hat{X}$ [productivity].

(T′.4) "$(L, X, Y) \in J'$, $L = 0$, *and* $Y \neq 0$" *implies* "$Y < \lambda X$" [importance of labor].

(T′.5) "$(L, X_1, Y_1) \in J'$, $(L, X_2, Y_2) \in J'$, $L > 0$, $X_1 \neq X_2$, *and* $0 < w < 1$" *implies that* "*there exists* $Y > wY_1 + (1 - w) Y_2$ *such that* $(L, wX_1 + (1 - w) X_2, Y) \in J'$" [weak strict convexity for outputs].

Let $Y_0 \in R^n_+$ be given; a feasible production program from the initial stock Y_0 is a pair of sequences $(\mathbf{X}, \mathbf{Y}) = (X_t, Y_t)_{t=0}^{\infty}$ such that

$$\begin{aligned} (L_t, X_t, Y_{t+1}) &\in J' \qquad \text{for all } t \in N \\ X_t &\leqslant Y_t \qquad \text{for all } t \in N. \end{aligned} \tag{2.2}$$

A feasible production program determines a consumption program $\mathbf{C} = (C_t)_{t=0}^{\infty}$ as

$$C_t = Y_t - X_t \qquad \text{for all } t \in N. \tag{2.3}$$

We refer to $\mathbf{X}$ (resp. $\mathbf{Y}$) as the input (resp. output) program.

II.c. *Reduction to per Capita Terms*

Define the *per capita variables*

$$x_t = X_t/L_t, \qquad y_t = Y_t/L_t, \qquad c_t = C_t/L_t \qquad \text{for } t \in N \tag{2.4}$$

using the assumption (T′.1). We can rewrite the feasibility conditions (2.2) as

$$(1, x_t, \lambda y_{t+1}) \in J', \qquad x_t \leqslant y_t \quad \text{for all } t \in N. \tag{2.5}$$

Define the set $J \subset R^n_+ \times R^n_+$ as $J = \{(x, y): (1, x, \lambda y) \in J'\}$. Then (2.5) is equivalent to "$(x_t, y_{t+1}) \in J$, $x_t \leqslant y_t$ for all $t \in N$." Also, from (2.3), one gets $c_t = y_t - x_t$ for all $t \in N$. We refer to the sequences $\mathbf{x} = (x_t)$, $\mathbf{y} = (y_t)$, $\mathbf{c} = (c_t)$ as *per capita* input, output, and consumption programs, respectively. For brevity, $(\mathbf{x}, \mathbf{y}, \mathbf{c})$ is called "a program." The set of all programs from an initial y_0 is denoted by $P(y_0)$. The following compactness result is known (see, e.g., Peleg [13, Lemma 4.1]):

R.1. *$P(y_0)$ is a convex subset of $s \times s \times s$ which is compact in the product topology; furthermore there is a positive real number $K(y_0)$ such that for all $(\mathbf{x}, \mathbf{y}, \mathbf{c}) \in P(y_0)$ one has*

$$\|y_t\| \leqslant K(y_0) \qquad \textit{for all } t \in N. \tag{2.6}$$

II.d. *The Optimality Criterion*

Alternative programs from y_0 are evaluated according to the utilities generated by consumptions. Let the one-period *felicity function* $u: R^n_+ \to R$ be assumed to satisfy:

(U.1) *u is continuous on R^n_+.*

(U.2) *u is strictly increasing* [i.e., $c_1 > c_2$ *implies* $u(c_1) > u(c_2)$] *on* $\{c \in R^n_+ : u(c) > u(0)\}$.

(U.3) *u is strictly concave on R^n_+.*

Given a *discount factor* γ $(0 < \gamma \leqslant 1)$, a program $(\bar{\mathbf{x}}, \bar{\mathbf{y}}, \bar{\mathbf{c}}) \in P(y_0)$ is *optimal* if

$$\lim_{T \to \infty} \sup \sum_{t=0}^{T} \gamma^t [u(c_t) - u(\bar{c}_t)] \leqslant 0 \tag{2.7}$$

for all programs $(\mathbf{x}, \mathbf{y}, \mathbf{c})$ in $P(y_0)$. We refer to the case $\gamma = 1$ (resp. $\gamma < 1$) as the *undiscounted* (*resp. discounted*) *case.*

II.e. *Competitive Programs*

Let $y_0 \in R^n_+$; a program $(\mathbf{x}, \mathbf{y}, \mathbf{c})$ in $P(y_0)$ is *competitive* if there exists a sequence $\mathbf{p} = (p_t)$ such that for all $t \in N$

$$\gamma^t u(c_t) - p_t c_t \geq \gamma^t u(c) - p_t c \qquad \text{for all } c \in R^n_+ \tag{2.8}$$

$$p_{t+1} y_{t+1} - p_t x_t \geq p_{t+1} y - p_t x \qquad \text{for all } (x, y) \in J. \tag{2.9}$$

In what follows, $\langle (\mathbf{x}, \mathbf{y}, \mathbf{c}); \mathbf{p} \rangle$ denotes a competitive program $(\mathbf{x}, \mathbf{y}, \mathbf{c})$ with its supporting price system $\mathbf{p}$.

III. Optimality of Competitive Programs: The Undiscounted Case

In Sections III and IV, we treat the *undiscounted case* $\gamma = 1$. In order to characterize competitive programs that are optimal we shall appeal to the following standard criterion (see, e.g., Peleg [13, Theorem 10.2]):

R.2. *Let* $y_0 \geq 0$; *if* $\langle (\mathbf{x}, \mathbf{y}, \mathbf{c}); \mathbf{p} \rangle$ *is a competitive program from* y_0 *such that the sequence* $\mathbf{p}$ *is bounded, then* $(\mathbf{x}, \mathbf{y}, \mathbf{c})$ *is optimal.*

Remark. The proof of R.2 uses the strict convexity assumption (T′.5).

We first recall some properties of golden rule programs in our model.

III.a. *Stationary Programs*

A program $(\mathbf{x}, \mathbf{y}, \mathbf{c})$ is *stationary* if $x_t = x_0$, $y_t = y_0$, and $c_t = c_0$ for all $t \in N$. Define

$$\begin{aligned} D &= \{(x, y): (x, y) \in J, y - x \geq 0\} \\ C &= \{c: c = y - x, (x, y) \in D\}. \end{aligned} \tag{3.1}$$

Note that both C and D are non-empty, compact, and convex. Define

$$u_* = \max\{u(c): c \in C\}. \tag{3.2}$$

A triplet (x, y, c) satisfying $(x, y) \in D$, $c \in C$ is a *stationary* triplet. Any such stationary triplet defines a stationary program in $P(y)$ by $x_t = x$, $y_t = y$, $c_t = c$ for all $t \in N$. Using the assumptions (T′.5) and (U.3) one can easily show that there is a *unique* stationary triplet (x^*, y^*, c^*) such that $u(c^*) = u_*$. We refer to (x^*, y^*, c^*) as the *golden rule triplet* and the stationary program $(\mathbf{x}^*, \mathbf{y}^*, \mathbf{c}^*)$ [that (x^*, y^*, c^*) defines] as the *golden* *rule program.* By (T′.3), there is some $(\hat{x}, \hat{y}) \in D$ such that $\hat{c} = \hat{y} - \hat{x} \gg 0$. Hence, by (U.2), $c^* \neq 0$ and this implies that $y^* > 0$. The following price-support property of the golden rule triplet will be used (see, e.g., Peleg [13, Lemma 6.6]):

R.3. *Let (x^*, y^*, c^*) be the golden rule triplet; there exists $p^* \in R^n_{++}$ such that*

$$u(c^*) - p^*c^* \geqslant u(c) - p^*c \qquad \textit{for all } c \in R^n_+ \tag{3.3}$$

$$p^*(y^* - x^*) \geqslant p^*(y - x) \qquad \textit{for all } (x, y) \in J. \tag{3.4}$$

Using (R.3), one shows that the stationary program $(\mathbf{x}^*, \mathbf{y}^*, \mathbf{c}^*)$ defined by the triplet (x^*, y^*, c^*) (i.e., the *golden rule program*) is competitive: the price system $\mathbf{p}^* = (p_t^*)$ supporting $(\mathbf{x}^*, \mathbf{y}^*, \mathbf{c}^*)$ is the stationary sequence $p_t^* = p^*$ (for all $t \in N$), where p^* satisfies (3.3) and (3.4).

Let (x^*, y^*, c^*) be the golden rule triplet with an associated p^* satisfying (3.3) and (3.4). We call (x^*, y^*, c^*, p^*) a *golden rule equilibrium.* Define a function $\delta^*: R^n_+ \to R_+$ as

$$\delta^*(c) = [u(c^*) - p^*c^*] - [u(c) - p^*c]. \tag{3.5}$$

One interprets δ^* as the *value loss* at c computed by using prices p^*. For any (x, y) in J, define a function $\pi^*: J \to R_+$ as

$$\pi^*(x, y) \equiv p^*(y^* - x^*) - p^*(y - x). \tag{3.6}$$

One interprets π^* as the *loss of intertemporal profit* at (x, y) computed as prices (p^*, p^*). We note the following "uniform value-loss" property:

R.4. *Let $\langle (x^*, y^*, c^*), p^* \rangle$ be the golden rule equilibrium.*

(a) *Suppose that M is a non-empty compact subset of R^n_+. Given any $\varepsilon > 0$, there is $\delta_1(\varepsilon) > 0$ such that "$\|c - c^*\| > \varepsilon$, $c \in M$" implies "$\delta^*(c) \geqslant \delta_1(\varepsilon)$."*

(b) *Let M' be a non-empty compact subset of J. Given $\varepsilon > 0$ there is $\delta_2(\varepsilon) > 0$ such that "$(x, y) \in M'$, $\|x - x^*\| > \varepsilon$" implies "$\pi^*(x, y) \geqslant \delta_2(\varepsilon)$."*

We now consider a competitive program $\langle (\bar{\mathbf{x}}, \bar{\mathbf{y}}, \bar{\mathbf{c}}); \bar{\mathbf{p}} \rangle$ from $y_0 > 0$. For each $t \in N$, define the function $\bar{\delta}_t: R^n_+ \to R_+$ as

$$\bar{\delta}_t(c) = [u(\bar{c}_t) - \bar{p}_t\bar{c}_t] - [u(c) - \bar{p}_t c]. \tag{3.7}$$

By (2.8), $\bar{\delta}_t(c) \geqslant 0$ for all $c \geqslant 0$. Also, for each $t \in N$, define the function $\bar{\pi}_t: J \to R_+$ as

$$\bar{\pi}_t(x, y) = [\bar{p}_{t+1}\bar{y}_{t+1} - \bar{p}_t\bar{x}_t] - [\bar{p}_{t+1}y - \bar{p}_t x]. \tag{3.8}$$

By (2.9), $\bar{\pi}_t(x, y) \geqslant 0$ for all $(x, y) \in J$. We now introduce the key element that enables us to signal optimality of competitive programs. For each $t \in N$, define

$$v_t \equiv (\bar{y}_t - y^*)(\bar{p}_t - p^*). \tag{3.9}$$

One can verify that

$$v_{t+1} - v_t = \bar{\pi}_t(x^*, y^*) + \pi^*(\bar{x}_t, \bar{y}_{t+1}) + \delta_t(c^*) + \delta^*(\bar{c}_t) \geqslant 0. \tag{3.10}$$

Before stating our first result on characterizing optimality of competitive programs in terms of the sequence (v_t), we introduce one more concept. The golden rule input $x^* > 0$ is said to be *expansible* if there is some $z \gg 0$ such that $(x^*, x^* + z) \in J$. Of course, if the golden rule triplet (x^*, y^*, c^*) is such that $c^* \gg 0$, x^* is necessarily expansible.

THEOREM 1. *Suppose that the golden rule stock x^* is expansible. Let $\langle (\bar{\mathbf{x}}, \bar{\mathbf{y}}, \bar{\mathbf{c}}); \bar{\mathbf{p}} \rangle$ be any competitive program from $\bar{y}_0 \geqslant 0$ such that*

$$v_t = (\bar{y}_t - y^*)(\bar{p}_t - p^*) \leqslant 0 \qquad \textit{for all } t \in N. \tag{3.11}$$

Then $(\bar{\mathbf{x}}, \bar{\mathbf{y}}, \bar{\mathbf{c}})$ is optimal.

Proof. By (R.2), it is enough to show that there is a constant $C > 0$ and a positive integer T such that $0 \leqslant \bar{p}_t \leqslant C$ for all $t \geqslant T$. We first claim that (3.11) implies

"*$\bar{c}_t$ converges to c^* as t goes to infinity.*" (3.12)

If the statement (3.12) is false, there are $\varepsilon' > 0$ and an infinite subset F of N such that $\|\bar{c}_t - c^*\| > \varepsilon'$ for all $t \in F$. Using (R.1) and (R.4)(a), one asserts the existence of some $\delta^1(\varepsilon') > 0$ such that $\delta^*(\bar{c}^t) \geqslant \delta^1(\varepsilon') > 0$ for all $t \in F$. But this means that $v_t > 0$ eventually in t (see (3.10)), a contradiction to (3.11). This argument establishes the claim (3.12). A similar argument relying upon (R.1) and (R.4)(b) establishes that

"*$\bar{x}_t$ converges to x^* as t goes to infinity.*" (3.13)

Since $\bar{y}_t = \bar{x}_t + \bar{c}_t$, we conclude that

"*$(\bar{x}_t, \bar{y}_t, \bar{c}_t)$ converges to (x^*, y^*, c^*) as t goes to infinity.*" (3.14)

If (p_t) is not bounded, choose a subsequence (retain same notation) such that $\|p_t\|/\|p_{t+1}\|$ is bounded, and $\|p_{t+1}\|$ goes to infinity. By (2.9), if we compare $(\bar{x}_t, \bar{y}_{t+1})$ with $(x^*, x^* + z)$, we get

$$\bar{p}_{t+1} \bar{y}_{t+1} - \bar{p}_t \bar{x}_t \geqslant \bar{p}_{t+1}(x^* + z) - \bar{p}_t x^*,$$

or

$$\frac{\bar{p}_{t+1}}{\|\bar{p}_{t+1}\|}\bar{y}_{t+1}-\frac{\bar{p}_t}{\|\bar{p}_{t+1}\|}\bar{x}_t \geqslant \frac{\bar{p}_{t+1}}{\|\bar{p}_{t+1}\|}(x^*+z)-\frac{\bar{p}_t}{\|\bar{p}_{t+1}\|}x^*. \tag{3.15}$$

Let $(\bar{p}, \bar{q})$ be a limit point of the sequence $(\bar{p}_{t+1}/\|\bar{p}_{t+1}\|, \bar{p}_t/\|\bar{p}_{t+1}\|)$. Clearly, $\|\bar{p}\|=1$, which means that $\bar{p}>0$. From (3.15), taking limits we get

$$\bar{p}y^*-\bar{q}x^* \geqslant \bar{p}x^*+\bar{p}z-\bar{q}x^*,$$

or

$$\bar{p}(x^*+c^*)-\bar{q}x^* \geqslant \bar{p}x^*+\bar{p}z-\bar{q}x^*,$$

or

$$\bar{p}c^* \geqslant \bar{p}z>0. \tag{3.16}$$

Now, by choosing $c=0$ on the right side of the inequality (2.8) we get (for all $t \in N$)

$$u(\bar{c}_{t+1})-\bar{p}_{t+1}\bar{c}_{t+1} \geqslant u(0),$$

or

$$\frac{u(\bar{c}_{t+1})}{\|p_{t+1}\|}-\frac{\bar{p}_{t+1}}{\|p_{t+1}\|}\bar{c}_{t+1} \geqslant \frac{u(0)}{\|\bar{p}_{t+1}\|}. \tag{3.17}$$

Since $\{u(\bar{c}_{t+1})\}$ is bounded and $\|p_{t+1}\|$ goes to infinity with t taking limits in (3.17) we get

$$\bar{p}c^* \leqslant 0. \tag{3.18}$$

The contradiction obtained from (3.16) and (3.18) establishes the theorem.
Q.E.D.

We now prove a converse to Theorem 1. Recall (from Gale [4]) that $(\mathbf{x}, \mathbf{y}, \mathbf{c}) \in P(y_0)$ is *good* if there exists a constant N_1 such that

$$\sum_{t=0}^{T}[u(c_t)-u_*] \geqslant N_1 \qquad \text{for all } T \geqslant 0. \tag{3.19}$$

The following properties related to good programs are needed:

R.5 (Gale [4]). *$(x, y, c) \in P(y_0)$ is either good or $\lim_{T\to\infty}\sum_{t=0}^{T}[u(c_t)-u_*]=-\infty$. For a good program $\lim_{T\to\infty}\sum_{t=0}^{T}[u(c_t)-u_*]$ exists and is finite, and* (i) $\lim_{t\to\infty}c_t=c^*$, (ii) $\lim_{t\to\infty}x_t=x^*$.

R.6 (Gale [4]). *If $y_0 \gg 0$, there is a good program in $P(y_0)$.*

We can now state and prove

THEOREM 2. *Let $\bar{y}_0 \gg 0$ and $(\bar{\mathbf{x}}, \bar{\mathbf{y}}, \bar{\mathbf{c}})$ be optimal in $P(\bar{y}_0)$. It is competitive at prices $\bar{p}_t$ and*

$$v_t = (\bar{y}_t - y^*)(\bar{p}_t - p^*) \leqslant 0 \qquad \textit{for all } t \in N. \tag{3.20}$$

Proof. For any $y \in R^n_+$, write

$$h(y) = \sup\left[\lim_{\tau \to \infty} \sum_{t=0}^{\tau} (u(c_t) - u_*): (\mathbf{x}, \mathbf{y}, \mathbf{c}) \in P(y)\right].$$

Since $\bar{y}_0 \gg 0$, (R.5) and (R.6) imply that $(\bar{x}, \bar{y}, \bar{c})$, which is optimal in $P(\bar{y}_0)$, is necessarily good. Hence, $h(\bar{y}_0)$ is finite and

$$h(\bar{y}_0) = \lim_{\tau \to \infty} \sum_{t=0}^{\tau} [u(\bar{c}_t) - u_*]. \tag{3.21}$$

Now, for any $T \geqslant 1$, the program $(\bar{x}_t, \bar{y}_t, \bar{c}_t)_{t=T}^{\infty}$ is optimal in $P(\bar{y}_T)$; it must also be a good program (otherwise, by (R.5), $\lim_{\tau \to \infty} \sum_{t=T}^{\tau} [u(\bar{c}_t) - u_*] = -\infty$, and this contradicts the fact that $(\bar{\mathbf{x}}, \bar{\mathbf{y}}, \bar{\mathbf{c}})$ is a good program from y_0). Hence, for any $T > 1$, one actually has

$$h(\bar{y}_T) = \lim_{\tau \to \infty} \sum_{t=T}^{\tau} [u(\bar{c}_t) - u_*]. \tag{3.22}$$

Recall that the golden rule program $(\mathbf{x}^*, \mathbf{y}^*, \mathbf{c}^*)$ defined by the triplet (x^*, y^*, c^*) is optimal in $P(y^*)$, and is competitive at (the stationary sequence of) prices $\mathbf{p}^* = (p^*)$. Clearly $h(y^*) = 0$. We now use the following result due to Peleg and Zilcha [15]:

R.7. *The optimal $(\mathbf{x}, \mathbf{y}, \mathbf{c})$ in $P(\bar{y}_0)$ is competitive at prices $\bar{\mathbf{p}} = (\bar{p}_t)$ and for all $\tau \in N$*

$$h(\bar{y}_\tau) - \bar{p}_\tau \bar{y}_\tau \geqslant h(y) - \bar{p}_\tau y \qquad \textit{for all } y \in R^n_+. \tag{3.23}$$

In particular, for all $\tau \in N$

$$h(\bar{y}_\tau) - \bar{p}_\tau \bar{y}_\tau \geqslant h(y^*) - \bar{p}_\tau y^* = -\bar{p}_\tau y^* \qquad [\text{recall that } h(y^*) = 0]. \tag{3.24}$$

Using (3.3) and (3.4) we can establish that for all $\tau \in N$,

$$p^* \bar{y}_0 - p^* y^* - p^*(\bar{x}_\tau - x^*) \geqslant \sum_{t=0}^{\tau} [u(\bar{c}_t) - u_*]. \tag{3.25}$$

Since $(\bar{\mathbf{x}}, \bar{\mathbf{y}}, \bar{\mathbf{c}})$ is a good program from $\bar{y}_0$, $\lim_{\tau \to \infty} \bar{x}_\tau = x^*$ (see (R.5)). Hence, taking limits in (3.25) as τ tends to infinity, and using (3.21), we get

$$p^* \bar{y}_0 - p^* y^* \geqslant h(\bar{y}_0). \tag{3.26}$$

Using the relation (3.22) and the remarks preceding it, one can apply the argument leading to (3.26) and derive, for *all* $\tau \in N$,

$$p^* \bar{y}_\tau - p^* y^* \geqslant h(\bar{y}_\tau). \tag{3.27}$$

By adding (3.24) and (3.27) we get (3.20).

IV. Remarks and Some Possible Extensions

(a). *The Discounted Case.* In the discounted case, a discount factor $\gamma \in (0, 1)$ is given. An analogue of Theorem 2 can be proved (under some "standard" assumptions on u and J') by using a very similar proof. Note that the existence of a "modified" golden rule equilibrium has been proved by Peleg and Ryder ([12], [14]) and the crucial separation argument leading to (R.7) can be completed by studying McKenzie [9] or Zilcha [18].

For deriving a result analogous to Theorem 1 in the discounted case, one possibility is to utilize the curvature conditions suggested by Rockafeller [17, p. 75]. We have not studied the implications of the Hamiltonian approach to the context of decentralization, but some direct calculations suggested that it was possible to obtain an analogue of Theorem 1 in a Hamiltonian framework.[2] A definitive treatment of the discounted case is now available in Dasgupta and Mitra [2].

(b). *Reachable Economies.* So far we have dealt with models that admit a golden rule equilibrium. The problem of characterizing optimality of competitive programs by means of conditions that are decentralizable in a reachable economy (see McFadden [8]) is very much open.[3]

(c). *Stochastic Versions.* We conjecture that, at least for the undiscounted case, our main results can be extended to multisector models where technological possibilities are influenced by random disturbances.[4]

[2] The relevant algebraic manipulations were spelled out in detail in an earlier version circulated as a working paper (No. 333) from the Department of Economics, Cornell University. We assumed that the Hamiltonian is α-convex in quantities and β-concave in current prices (see Rockafeller [17, p. 75] for a precise definition). Interestingly enough, while our proof of Theorem 1 in the undiscounted case exploits the turnpike property (3.14), the arguments in the discounted case did not impose restrictions on γ typically needed to obtain turnpike results ("γ close to one").

[3] Proress in this direction also has been made by Professors Dasgupta and Mitra [3].

[4] See the paper by Nyarko [10], which confirms our conjecture.

(d). An interesting observation on the role of (U.3) has been made by professor Tapan Mitra. Suppose that the *strict concavity* of u is replaced by *Concavity* [i.e., (U.3) is replaced by a weaker (U.3)′: "u is concave"]. By using the weak strict convexity assumption (T′.5) on the technology, one is still able to obtain (R.4)(b) which, in turn, leads to (3.13). A modification of the subsequent sequential compactness argument in Theorem 1 leads to the same conclusion [replace c^* by an appropriate limit point of the subsequence (c_{t+1})]. Hence, *strict* concavity of the felicity function is not needed in the proof of Theorem 1. Of course, without (U.3), uniqueness of the golden rule triplet and results like (R.4)(a) and (3.12) do not hold: but these are not essential to our main point on decentralization.

(e). A referee has rightly pointed out that it will be useful to have an example showing that the assumption on expansibility of x^* [appearing in Theorem 1] is indispensable [or to have a proof that disposes of this assumption].

References

1. D. Cass and M. Majumdar, Efficiency, consumption value maximization and capital value transversality: A unified view, *in* "Equilibrium, Value and Growth, Essays in Honor of Lionel McKenzie" (Green and Scheinkman, eds.), Academic Press, New York, 1979.
2. S. Dasgupta and T. Mitra, Characterization of intertemporal optimality in terms of decentralizable conditions: The discounted case, *J. Econ. Theory* **45** (1988), 274–287.
3. S. Dasgupta and T. Mitra, Intertemporal optimality in a closed linear model of production, *J. Econ. Theory* **45** (1988), 288–315.
4. D. Gale, On optimal development in a multi-sector economy, *Rev. Econ. Stud.* **34** (1967), 1–18.
5. L. Hurwicz, On informationally decentralized systems, *in* "Decision and Organization" (McGuire and Radner, eds.), pp. 297–336, North-Holland, Amsterdam, 1972.
6. L. Hurwicz and M. Majumdar, Optimal intertemporal allocation mechanisms and decentralization of decisions, *J. Econ. Theory* **45** (1988), 228–261.
7. E. Malinvaud, Capital accumulation and efficient allocation of resources, *Econometrica* **21** (1953), 233–268.
8. D. McFadden, The evaluation of development programs, *Rev. Econ. Stud.* **34** (1967), 25–50.
9. L. McKenzie, Turnpike theory, *Econometrica* **44** (1976), 841–866.
10. Y. Nyarko, On characterizing optimality of stochastic competitive processes, *J. Econ. Theory* **45** (1988), 316–329.
11. H. Nikaido, "Convex Structures and Economic Theory," Academic Press, New York, 1968.
12. B. Peleg and H. Ryder, On optimal consumption plans in a multisector economy, *Rev. Econ. Stud.* **39** (1972), 159–169.
13. B. Peleg, On competitive prices for optimal consumption plans, *SIAM J. Appl. Math.* **26** (1974), 239–53.
14. B. Peleg and H. Ryder, The modified golden rule of a multisector economy, *J. Math. Econ.* **1** (1974), 193–198.

15. B. PELEG AND I. ZILCHA, On competitive prices for optimal consumption plans, II, *SIAM J. Appl. Math.* **32** (1977), 127–131.
16. F. RAMSEY, A mathematical theory of savings, *Econ. J.* **38** (1928), 543–559.
17. R. T. ROCKAFELLER, Saddle points of hamiltonian systems in convex lagrange problems having a non-zero discount rate, *J. Econ. Theory* **12** (1976), 71–114.
18. I. ZILCHA, Characterization by prices of optimal programs under uncertainty, *J. Math. Econ.* **3** (1976), 173–183.

4

Characterization of Intertemporal Optimality in Terms of Decentralizable Conditions: The Discounted Case

SWAPAN DASGUPTA*

Department of Economics, Dalhousie University, Halifax, Nova Scotia B3H 3J5, Canada

AND

TAPAN MITRA

Department of Economics, Cornell University, Ithaca, New York 14853

The paper studies the problem of characterizing the optimality of competitive programs in terms of "decentralizable" conditions. We show that, when future utilities are discounted, and the optimal stationary stock is proportionately expansible, then optimality of competitive programs can be characterized by the condition that the scalar product of the difference of prices and quantities, between those of the given competitive program and those of the optimal stationary program, be non-positive period by period. *Jornal of Economic Literature* Classification Number: 111.

1. INTRODUCTION

The aim of this paper is to present some results on the characterization of optimality of "competitive" programs, in terms of a "decentralizable" condition, in the context of a standard Ramsey-type multisector growth model, where the technology, the period welfare function, and the period discount factor, which is assumed *less than one*, are stationary over time.

It is well known [from the price characterization results of Cass and Majumdar [2], Peleg [8, 9], Peleg and Ryder [10], Peleg and Zilcha [12], and Weitzman [13]] that in (discounted and undiscounted)

* We are indebted to Mukul Majumdar for introducing us to the problems of intertemporal decentralization and for extremely helpful conversations. We thank M. Berliant, D. Cass, L. W. McKenzie, P. Romer, K. Shell, and a referee for many useful comments. Research of the second author was supported by a National Science Foundation grant. Research on this paper was started while the first author was on sabbatic leave at Cornell University and Instituto Torcuato Di Tella, Buenos Aires.

multisectoral growth models, a program, from positive initial stocks, is optimal if and only if there exists an associated sequence of prices such that (a) these prices "support" the welfare function and the technology, at the consumption and the input–output vectors respectively, in each period, along the given program; and (b) the value of inputs at these prices converges to zero along the program (this being the relevant condition in the "discounted" case, that is, when the discount factor is less than one; in the "undiscounted" case, that is, when the discount factor equals one, the relevant condition is that the input values be uniformly bounded along the program). Condition (a) above is typically interpreted as maximization of profits and maximization of welfare (subject to an appropriate budget constraint) at the input–output vector and the consumption vector, respectively, period by period, along the given program (see Gale and Sutherland [5]).

Recently Brock and Majumdar [1] have investigated the possibility of characterizing the optimality of competitive programs in terms of conditions which can be verified by agents in an "informationally decentralized" mechanism. For a detailed discussion of the problem, we refer the reader to their paper. For our purpose here, it is sufficient to note that their objective is to replace the transversality condition, [condition, (b) above], in characterizing optimality of competitive programs, by a condition which can be verified by the agents, period by period, on the basis of information regarding prices in each period and possibly some additional fixed information or finite "messages" which are transmitted in each period.

Brock and Majumdar [1] show that in models, where there exists an "optimal stationary program" (abbreviated as o.s.p.), which has a "stationary price support," a characterization of optimality of competitive programs along such lines is possible. More specifically, they show that *in the undiscounted case*, if the input stock along the o.s.p. is "expansible" (that is, an output vector can be produced from such a stock, which provides more of each good than was provided in the initial stock), then the transversality condition (b) can be replaced by the condition that (c) the scalar product of the difference of prices and of quantities, between those of the given competitive program and those of the o.s.p., be nonpositive period by period.

The main result of our paper is to show that, *in the discounted case*, optimality of competitive programs can be characterized by (the natural analog) of the condition proposed by Brock and Majumdar. More precisely, we show that [when the discount factor is less than one], and the input stock along the o.s.p. is "proportionately expansible" [which is a slightly weaker requirement than being "expansible"], then the transversality condition (b) can be replaced by the "decentralizable condition" (c) in characterizing optimality of competitive programs.

It should be noted at this point that in the usual characterization results, it is well known that the transversality condition can be replaced by the condition that (b′) the prices "support" the value functon in each period (see Proposition 2.3 below). It is also known that (b′) implies (c) (see Theorem 2.2 below), so that the novel aspect of the new characterization results is that condition (c) is *sufficient* to guarantee that a competitive program is optimal (Theorem 3.1).

It may be of interest to observe a difference in the nature of the characterization results involving the transversality condition and those involving condition (c). Using the transversality condition, optimality or non-optimality can be verified, loosely speaking, only "at infinity," that is, by investigating the *asymptotic* behaviour of input value along a competitive program. In contrast, using condition (c), non-optimality (but not optimality) can always be detected within *some finite* horizon. A "price" is paid for this "gain." In the characterization involving the transversality condition, only knowledge regarding the competitive program is required, that is, the quantities as well as the supporting prices along the program. In contrast, in the characterization results in this paper involving condition (c), knowledge is required, in addition, of the quantities and prices of the o.s.p. Furthermore, it seems that typically, some mild "regularity" assumptions need to be made regarding the o.s.p.

Finally, it may be worthwhile to note that, like the characterization results involving condition (c), the standard characterizations using (b′) also show that non-optimality can always be detected within some finite horizon. Here, however, the difference in the two types of results lies in the extent and nature of the *information* required in the conditions (b′) and (c), respectively. In (b′), knowledge is required effectively of the entire value function, whereas in (c), the agents need to know only a finite set of numbers, namely the constant output and the constant "current prices" of the o.s.p. Furthermore, the knowledge of the value function is just a step away from a parametric solution of the intertemporal optimization problem, namely the optimal policy function. In contrast, (c) requires the knowledge of the optimal policy function at a single point, namely the initial stock of the o.s.p.

2. Preliminaries

2a. *Notation*

R^n denotes the n-dimensional real vector space of n-tuples of real numbers. For x, y in R^n, $x \geqslant y$ means $x_i \geqslant y_i$ for $i = 1, \ldots, n$; $x > y$ means $x \geqslant y$ and $x \neq y$; $x \gg y$ means $x_i > y_i$ for $i = 1, \ldots, n$. R^n_+ denotes the set $\{x$ in

$R^n: x \geqslant 0\}$, and R^n_{++} denotes the set $\{x \text{ in } R^n: x \gg 0\}$. For x in R^n the sum norm of x (denoted by $\|x\|$) is defined by $\|x\| = (\sum_{i=1}^n |x_i|)$. We denote the vector $(1, 1, ..., 1)$ in R^n by e.

2b. *The Model*

The model is described by a triplet (Ω, w, δ), where Ω, a subset of $R^n_+ \times R^n_+$, is the *technology set*, $w: R^n_+ \to R$ is the period *walfare function*, and δ is the *discount factor* satisfying $0 < \delta < 1$. Points in Ω are written as an ordered pair (x, y), where x stands for the (initial) stock of inputs and y stands for the (final) output which can be produced with inputs x.

We shall need the following assumptions on Ω and w.

(A.1) (a) $(0, 0)$ *is in* Ω; (b) $(0, y)$ *is in* Ω *implies* $y = 0$.

(A.2) Ω *is closed.*

(A.3) *There exists a number* $\beta_0 > 0$ *such that, if* (x, y) *is in* Ω *and* $\|x\| \geqslant \beta_0$, *then* $\|y\| \leqslant \|x\|$.

(A.4) *If* (x, y) *is in* Ω, $x' \geqslant x$, *and* $0 \leqslant y' \leqslant y$ *then* (x', y') *is in* Ω.

(A.5) Ω *is convex.*

(A.6) w *is continuous.*

(A.7) *If* c, c' *are in* R^n_+ *and* $c \geqslant c'$, *then* $w(c) \geqslant w(c')$; $w(e) > w(0)$.

(A.8) w *is concave.*

2c. *Programs*

A *program* from $\tilde{y}$ in R^n_+ is a sequence $\langle x(t), y(t)\rangle$ such that

$$y(0) = \tilde{y}; \quad 0 \leqslant x(t) \leqslant y(t) \quad \text{and} \quad (x(t), y(t+1)) \text{ is in } \Omega \quad \text{for} \quad t \geqslant 0.$$

Associated with a program $\langle x(t), y(t)\rangle$ from $\tilde{y}$ is a *consumption sequence* $\langle c(t)\rangle$ defined by

$$c(t) = y(t) - x(t) \quad \text{for} \quad t \geqslant 0.$$

To proceed further, we need the familiar preliminary result that programs from $\tilde{y}$ are uniformly bounded by a number which depends only on $\tilde{y}$ and β_0. We first establish

LEMMA 2.1. *Under* (A.3) *and* (A.4), *if* (x, y) *is in* Ω, *then*

$$\text{(i)} \quad \|x\| \leqslant \beta_0 \textit{ implies } \|y\| \leqslant \beta_0; \qquad \text{(ii)} \quad \|y\| \leqslant \text{Max}\{\|x\|, \beta_0\}.$$

Proof. (i) Suppose on the contrary there exists (x^0, y^0) in Ω satisfying $\|x^0\| \leqslant \beta_0$ and $\|y^0\| > \beta_0$. Define $x' \equiv x^0 + [\beta_0 - \|x^0\|](e/n)$. Since $\beta_0 \geqslant \|x^0\|$, therefore, $x' \geqslant x^0$. Hence, by (A.4), (x', y^0) is in Ω. But

$\|x'\| = \beta_0$ and hence, $\|y^0\| > \|x'\|$. This contradicts (A.3) and, therefore, establishes (i).

(ii) If $\|x\| \leqslant \beta_0$, then by (i), $\|y\| \leqslant \beta_0 \leqslant \text{Max}\{\|x\|, \beta_0\}$. If $\|x\| \geqslant \beta_0$ then, by (A.3), $\|y\| \leqslant \|x\| \leqslant \text{Max}\{\|x\|, \beta_0\}$. This establishes (ii). ∎

LEMMA 2.2. *Let $\tilde{y}$ in R^n_+ be given. Define $B \equiv \text{Max}\{\|\tilde{y}\|, \beta_0\}$. Under* (A.3) *and* (A.4), *if $\langle x(t), y(t)\rangle$ is a program from $\tilde{y}$ then $(\|x(t)\|, \|y(t)\|, \|c(t)\|) \leqslant (B, B, B)$, for $t \geqslant 0$.*

Proof. Since $0 \leqslant c(t) = y(t) - x(t) \leqslant y(t)$ for $t \geqslant 0$, therefore, we only need to show that

$$(\|x(t)\|, \|y(t)\|) \leqslant (B, B) \qquad \text{for} \quad t \geqslant 0. \tag{2.1}$$

First, $\|x(0)\| \leqslant \|y(0)\| = \|\tilde{y}\| \leqslant B$. Hence, (2.1) holds for $t = 0$. Consider any integer $\tau > 0$. Suppose (2.1) holds for $t = \tau$. Then $\|x(\tau+1)\| \leqslant \|y(\tau+1)\| \leqslant \text{Max}\{\|x(\tau)\|, \beta_0\}$ (using Lemma 2.1(ii)) $\leqslant B$ (since $\|x(\tau)\| \leqslant B$ by hypothesis). Thus, (2.1) holds for $t = \tau + 1$. This establishes (2.1) by induction and, therefore, the lemma. ∎

2d. *Optimal and Competitive Programs*

In view of Lemma 2.2 and continuity of w it is clear that for every program $\langle x(t), y(t)\rangle$ from $\tilde{y}$, $\sum_{t=0}^{\infty} \delta^t w(c(t))$ is absolutely convergent. We may, therefore, make the following definition: a program $\langle x^0(t), y^0(t)\rangle$ from $\tilde{y}$ is an *optimal program* if, for every program $\langle x(t), y(t)\rangle$ from $\tilde{y}$, we have

$$\sum_{t=0}^{\infty} \delta^t w(c^0(t)) \geqslant \sum_{t=0}^{\infty} \delta^t w(c(t)).$$

The following lemma is a standard consequence of Lemma 2.2 and is stated without proof.

LEMMA 2.3. *Under* (A.1)–(A.4), *and* (A.6), *if $\tilde{y}$ is in R^n_+ then there is an optimal program $\langle x(t), y(t)\rangle$ from $\tilde{y}$.*

We may then define the *value function* $V\colon R^n_+ \to R$ by $V(y) = \sum_{t=0}^{\infty} \delta^t w(c(t))$, where $\langle c(t)\rangle$ is the consumption sequence associated with some optimal program $\langle x(t), y(t)\rangle$ from y.

A sequence $\langle x(t), y(t), p(t)\rangle$ from $\tilde{y}$ is a *competitive program* if $\langle x(t), y(t)\rangle$ is a program from $\tilde{y}$, $p(t)$ is in R^n_+ for $t \geqslant 0$, and

$$\delta^t w(c(t)) - p(t)\, c(t) \geqslant \delta^t w(c) - p(t)\, c \qquad \text{for} \quad c \text{ in } R^n_+, \quad t \geqslant 0 \tag{2.2}$$

and

$$p(t+1)\,y(t+1)-p(t)\,x(t) \geqslant p(t+1)\,y-p(t)\,x \qquad \text{for} \quad (x, y) \text{ in } \Omega, \quad t \geqslant 0. \tag{2.3}$$

Adding up the inequalities (2.2) and (2.3), we note that if $\langle x(t), y(t), p(t)\rangle$ is a competitive program, then

$$\delta^t w(c(t))+p(t+1)\,y(t+1)-p(t)\,y(t) \geqslant \delta^t w(c)+p(t+1)\,y-p(t)(x+c) \tag{2.4}$$

for all (x, y) in Ω and all c in R^n_+.

A competitive program $\langle x(t), y(t), p(t)\rangle$ is said to satisfy the *transversality condition* if

$$\lim_{t \to \infty} p(t)\,x(t)=0. \tag{2.5}$$

2e. *Characteriation of Optimality of Competitive Programs in Terms of a Transversality Condition*

It is well known that a competitive program satisfying the transversality condition is optimal. We state this in Proposition 2.1 below for ready reference. The proof of this proposition is standard and therefore omitted. It need only be noted that (A.3), (A.4), and (A.6) guarantee (by virtue of Lemma 2.2) absolute convergence of $\sum_{t=0}^{\infty} \delta^t w(c(t))$ for any program $\langle x(t), y(t)\rangle$ from $\tilde{y}$ in R^n_+.

PROPOSITION 2.1. *Under* (A.3), (A.4), *and* (A.6), *if* $\langle x(t), y(t), p(t)\rangle$ *is a competitive program from* $\tilde{y}$ *in* R^n_+ *and*

$$\liminf_{t \to \infty} p(t)\,x(t)=0 \tag{2.6}$$

then $\langle x(t), y(t)\rangle$ *is an optimal program from* $\tilde{y}$.

The converse of the above proposition (Proposition 2.2 below) requires the use of the convex structure of the model. A version of it has been established [under somewhat different assumptions than the ones we use] by Peleg [8] and Peleg and Ryder [10]. The version we report here can be obtained from the result of Weitzman [13], and is proved in Dasgupta and Mitra [4].

A vector x in R^n_+ is *sufficient* if there is y in R^n_{++} such that (x, y) is in Ω. We shall need

(A.9) *There exists a sufficient vector in* R^n_+.

PROPOSITION 2.2. *Under* (A.1)–(A.9) *if* $\langle x(t), y(t)\rangle$ *is an optimal* *program from $\tilde{y}$ in R^n_{++}, then there exists a price sequence $\langle p(t) \rangle$ in R^n_+ such that*

(i) $\langle x(t), y(t), p(t) \rangle$ *is competitive,*

(ii) $\delta^t V(y(t)) - p(t)\, y(t) \geqslant \delta^t V(y) - p(t)\, y \qquad$ *for y in R^n_+, $t \geqslant 0$,* (2.7)

and

(iii) $\lim_{t \to \infty} p(t)\, x(t) = 0.$ (2.8)

Conditions (2.7) and (2.8) in the above result are not "independent." For a competitive program, (2.7) implies (2.8), and (2.8) implies (2.7). We note this formally in the next result.

PROPOSITION 2.3. *Under* (A.1)–(A.4) *and* (A.6), *if $\langle x(t), y(t), p(t) \rangle$ is a competitive program from $\tilde{y}$ in R^n_+, then it satisfies* (2.7) *if and only if it satisfies* (2.8).

Proof. If $\langle x(t), y(t), p(t) \rangle$ is competitive and satisfies (2.7), then $\delta^t V(y(t)) - p(t)\, y(t) \geqslant \delta^t V(0) - p(t)\, 0$. Hence $0 \leqslant p(t)\, y(t) \leqslant \delta^t [V(y(t)) - V(0)]$. Now, $V(y(t))$ is bounded above (by Lemma 2.2 and (A.6)), and V is defined over R^n_+ (by Lemma 2.3). Consequently, we have $\lim_{t \to \infty} p(t)\, y(t) = 0$, which implies (2.8), since $0 \leqslant x(t) \leqslant y(t)$ and $p(t) \geqslant 0$.

If $\langle x(t), y(t), p(t) \rangle$ is competitive and satisfies (2.8), then we verify (2.7) as follows. Pick any $T \geqslant 0$ and y in R^n_+. Let $\langle x'(s), y'(s) \rangle$ be any program from y. Since $\langle x(t), y(t), p(t) \rangle$ is competitive, so by (2.4), we have for $t \geqslant T$,

$$\delta^t [w(c'(t-T)) - w(c(t))] \leqslant [p(t+1)\, y(t+1) - p(t)\, y(t)] - [p(t+1)\, y'(t-T+1) - p(t)\, y'(t-T)].$$

Summing this from $t = T$ to $t = T+N$, we have

$$\begin{aligned} &\sum_{T}^{T+N} \delta^t [w(c'(t-T)) - w(c(t))] \\ &\quad \leqslant [p(T+N+1)\, y(T+N+1) - p(T)\, y(T)] \\ &\qquad - [p(T+N+1)\, y'(N+1) - p(T)\, y'(0)] \\ &\quad \leqslant p(T+N+1)\, y(T+N+1) - p(T)\, y(T) + p(T)\, y'(0). \end{aligned}$$

Thus, we have

$$\delta^T \left[\sum_{0}^{N} \delta^s w(c'(s)) - \sum_{T}^{T+N} \delta^{t-T} w(c(t)) \right] \leqslant p(T+N+1)\, y(T+N+1) - p(T)\, y(T) + p(T)\, y.$$

Since the sums in the above expressions converge as $N \to \infty$ [by (A.1)–(A.4) and (A.6)], and (2.8) holds, so

$$\delta^T \left[\sum_0^{\infty} \delta^s w(c'(s)) - \sum_T^{\infty} \delta^{t-T} w(c(t)) \right] \leqslant p(T)\, y - p(T)\, y(T).$$

By the principle of optimality, $\langle x(T+s), y(T+s) \rangle$ is an optimal program from $y(T)$, and so

$$\sum_T^{\infty} \delta^{t-T} w(c(t)) = \sum_0^{\infty} \delta^s w(c(T+s)) = V(y(T)).$$

Consequently we have

$$\delta^T V(y) - \delta^T V(y(T)) \leqslant p(T)\, y - p(T)\, y(T),$$

which establishes the inequality in (2.7) for the given y and T. Since y in R^n_+ and $T \geqslant 0$ were arbitrary, so (2.7) is satisfied. ∎

2f. *An Optimal Stationary Program*

A program $\langle x^*(t), y^*(t) \rangle$ from y^* is a *stationary program* if there is x^* such that $(x^*(t), y^*(t)) = (x^*, y^*)$ for $t \geqslant 0$. It is an *optimal stationary program* (abbreviated as o.s.p.) if it is a stationary program and it is optimal from y^*. In this case, we refer to x^* as an *optimal stationary stock* (abbreviated as o.s.s.). We refer to an o.s.p. as $\langle x^*, y^* \rangle$ with obvious interpretation, and its associated consumption sequence as $\langle c^* \rangle$, where $c^* \equiv y^* - x^*$. A vector q^* in R^n_+ is a *stationary price support* for an o.s.p. $\langle x^*, y^* \rangle$ if $\langle x^*(t), y^*(t), p^*(t) \rangle \equiv \langle x^*, y^*, \delta^t q^* \rangle$ is competitive from y^*.

The question of existence of an o.s.p. has been discussed extensively in the literature, most recently by Khan and Mitra [6] and McKenzie [7]. We note below a proposition on the existence of an optimal stationary program *with stationary price support.* A version of this result has been established by Peleg and Ryder [11], under somewhat different assumptions than the ones we use. The version we report here can be obtained from the result of Khan and Mitra [6], and is proved in Dasgupta and Mitra [4].

The technology set Ω is called *δ-productive* if there exists $(\hat{x}, \hat{y})$ in Ω, such that $\delta \hat{y} \gg \hat{x}$. We shall need

(A.10) *Ω is δ-productive.*

PROPOSITION 2.4. *Under* (A.1)–(A.8) *and* (A.10), *there is* (x^*, y^*) *in* Ω, $c^* \equiv y^* - x^* > 0$, $w(c^*) > w(0)$, *and* q^* *in* R^n_+ *such that* $\langle x^*, y^* \rangle$ *is an o.s.p. with stationary price support,* q^*.

For the remainder of the paper we fix one particular o.s.p. $\langle x^*, y^* \rangle$ and its stationary price support q^* for the purpose of definitions to follow. Furthermore, we denote $\delta^t q^*$ by $p^*(t)$ for $t \geqslant 0$.

3. Characterization of Optimality of Competitive Programs in Terms of a Decentralizable Condition

In this section, we show that optimality of competitive programs can be characterized in terms of the simple decentralizable rule of Brock and Majumdar [1], which requires for its verification in each period knowledge of current prices (that is, measured in terms of current welfare) and output quantities and knowledge of (x^*, y^*, q^*). Clearly, the difficult part of this characterization is to show that if a competitive program satisfies the decentralizable rule [that the scalar product of the difference of prices and quantities, between those of the given competitive program and those of the o.s.p., be non-positive, period by period] then it is optimal. We first establish two preliminary results (Lemmas 3.1 and 3.2) and then provide the main theorem of the paper (Theorem 3.1).

If $\langle x(t), y(t), p(t) \rangle$ is a competitive program from $\tilde{y}$ in R^n_+, then we denote

$$\mu(t) \equiv (p(t) - p^*(t))\,(x(t) - x^*(t)) \qquad \text{for} \quad t \geqslant 0$$

$$\nu(t) \equiv (p(t) - p^*(t))\,(y(t) - y^*(t)) \qquad \text{for} \quad t \geqslant 0$$

$$\theta(t) \equiv \mu(t+1) - \mu(t) \qquad \text{for} \quad t \geqslant 0.$$

Furthermore, we denote the *current price sequence* associated with it, $\langle p(t)/\delta^t \rangle$, by $\langle q(t) \rangle$.

Lemma 3.1. *Suppose $\langle x(t), y(t), p(t) \rangle$ is a competitive program from $\tilde{y}$. Then the following conditions hold:*

(i) $\nu(t+1) \geqslant \mu(t) \geqslant \nu(t)$ *for* $t \geqslant 0$,

(ii) $\mu(t+1) \geqslant \mu(t)$ *for* $t \geqslant 0$.

Proof. Since $\langle x(t), y(t), p(t) \rangle$ is competitive, so by (2.2), we have for $t \geqslant 0$,

$$\delta^t [w(c(t)) - w(c^*)] \geqslant p(t)[c(t) - c^*]. \tag{3.1}$$

Since $\langle x^*, y^*, p^*(t) \rangle$ is competitive, so by (2.2), we have for $t \geqslant 0$,

$$\delta^t [w(c^*) - w(c(t))] \geqslant -p^*(t)[c(t) - c^*]. \tag{3.2}$$

Adding (3.1) and (3.2), we get

$$[p(t)-p^*(t)][c(t)-c^*]\leqslant 0 \qquad \text{for} \quad t\geqslant 0,$$

so that

$$[p(t)-p^*(t)][x(t)-x^*]\geqslant[p(t)-p^*(t)][y(t)-y^*] \qquad \text{for} \quad t\geqslant 0. \quad (3.3)$$

Similarly, since $\langle x(t), y(t), p(t)\rangle$ is competitive, so by (2.3), we have for $t\geqslant 0$,

$$p(t+1)[y(t+1)-y^*]\geqslant p(t)[x(t)-x^*]. \quad (3.4)$$

And, since $\langle x^*, y^*, p^*(t)\rangle$ is competitive, so by (2.3), we have for $t\geqslant 0$,

$$-p^*(t+1)[y(t+1)-y^*]\geqslant -p^*(t)[x(t)-x^*]. \quad (3.5)$$

Adding (3.4) and (3.5), we get for $t\geqslant 0$,

$$[p(t+1)-p^*(t+1)][y(t+1)-y^*]\geqslant[p(t)-p^*(t)][x(t)-x^*]. \quad (3.6)$$

Combining (3.3) and (3.6) yields (i).

To establish (ii), note simply from (i) that $\mu(t+1)\geqslant v(t+1)$ for $t\geqslant 0$, and $v(t+1)\geqslant\mu(t)$ for $t\geqslant 0$, so that the conclusion is obvious. ∎

LEMMA 3.2. *Suppose $\langle x(t), y(t), p(t)\rangle$ is a competitive program from $\tilde{y}$, which satisfies $\mu(t)\leqslant 0$ for $t\geqslant 0$. Then, the following conditions hold*:

(i) $\theta(t)\geqslant 0$ *for* $t\geqslant 0$, *and* $\sum_0^\infty \theta(t)<\infty$,

(ii) $\theta(t)\to 0$ *as* $t\to\infty$.

Proof. From Lemma 3.1(ii), we have $\theta(t)\geqslant 0$ for $t\geqslant 0$. Also, for $T\geqslant 0$,

$$S(T)=\sum_0^T\theta(t)=\sum_0^T[\mu(t+1)-\mu(t)]=\mu(T+1)-\mu(0)\leqslant -\mu(0)$$

since $\mu(T+1)\leqslant 0$, by hypothesis. Thus, $\langle S(T)\rangle$ is a monotonically non-decreasing sequence, bounded above, hence it converges. This establishes (i).

It follows directly from (i) that $\theta(t)\to 0$ *as* $t\to\infty$, which is (ii). ∎

A vector x in R^n_+ is called *expansible* if there is $y\gg x$, such that (x, y) is in Ω. It is called *proportionately expansible* if there is $\lambda>1$, such that $(x, \lambda x)$ is in Ω. Clearly, if x is expansible, it is also proportionately expansible; the converse is false.

THEOREM 3.1. *Suppose x^* is proportionately expansible. If $\langle x(t), y(t), p(t)\rangle$ is a competitive program from $\tilde{y}$, which satisfies*

$$(q(t)-q^*)(y(t)-y^*)\leqslant 0 \qquad \textit{for} \quad t\geqslant 0, \tag{3.7}$$

then $\langle x(t), y(t)\rangle$ is an optimal program from $\tilde{y}$.

Proof. We note right at the outset that (3.7) implies that $v(t)\leqslant 0$ for $t\geqslant 0$. This, in turn, implies that $\mu(t)\leqslant 0$ for $t\geqslant 0$ by Lemma 3.1, and so $\theta(t)\to 0$ as $t\to\infty$ by Lemma 3.2.

Our first task now is to show that

$$p(t)\,x^*\to 0 \qquad \text{as} \quad t\to\infty. \tag{3.8}$$

To this end, use the competitive conditions for $\langle x(t), y(t), p(t)\rangle$ to write for $t\geqslant 0$,

$$\begin{aligned}&\delta^{t+1}w(c(t+1))+p(t+1)\,x(t+1)-p(t)\,x(t)\\&\quad\geqslant\delta^{t+1}w(c)+p(t+1)(y-c)-p(t)\,x\end{aligned} \tag{3.9}$$

for all (x, y) in Ω, and all c in R^n_+. Since x^* is proportionately expansible, there is $\lambda>1$ such that $(x^*, \lambda x^*)$ is in Ω. Using this in (3.9), we get for $t\geqslant 0$,

$$\begin{aligned}&\delta^{t+1}w(c(t+1))+p(t+1)\,x(t+1)-p(t)\,x(t)\\&\quad\geqslant\delta^{t+1}w(0)+p(t+1)\,\lambda x^*-p(t)\,x^*.\end{aligned} \tag{3.10}$$

Transposing terms in (3.10), we get for $t\geqslant 0$

$$\begin{aligned}&\delta^{t+1}[w(c(t+1))-w(0)]+p(t+1)[x(t+1)-x^*]-p(t)[x(t)-x^*]\\&\quad\geqslant(\lambda-1)\,p(t+1)\,x^*.\end{aligned} \tag{3.11}$$

Note that $p(t+1)[x(t+1)-x^*]=[p(t+1)-p^*(t+1)][x(t+1)-x^*]+p^*(t+1)[x(t+1)-x^*] = \mu(t+1)+p^*(t+1)[x(t+1)-x^*]$. Similarly, $p(t)[x(t)-x^*]=\mu(t)+p^*(t)[x(t)-x^*]$. Using these in (3.11), we get

$$\begin{aligned}&\delta^{t+1}[w(c(t+1))-w(0)]+\mu(t+1)+p^*(t+1)[x(t+1)-x^*]\\&\quad-\mu(t)-p^*(t)[x(t)-x^*]\geqslant(\lambda-1)\,p(t+1)\,x^*.\end{aligned} \tag{3.12}$$

Looking at the left-hand side of (3.12), we note that $[\mu(t+1)-\mu(t)]\equiv\theta(t)$ converges to zero as $t\to\infty$. The term $p^*(t)[x(t)-x^*]$ clearly converges to zero, since $p^*(t)=\delta^t q^*\to 0$ as $t\to\infty$, and $\|x(t)\|\leqslant B$; the same observation holds for the term $p^*(t+1)\,[x(t+1)-x^*]$. Finally, $\|c(t+1)\|\leqslant B$ and continuity of w [by (A.6)] imply that $w(c(t+1))$ is bounded above, while $w(0)$ is in R; so $\delta^{t+1}[w(c(t+1))-w(0)]\to 0$ as $t\to\infty$, since $\delta^{t+1}\to 0$ as $t\to\infty$. Thus, (3.8) follows from (3.12).

Our second task is to show that

$$p(t)\,x(t)\to 0 \quad \text{as} \quad t\to\infty. \tag{3.13}$$

This is rather easy, given (3.8). To see this, note that since $\mu(t)\leqslant 0$ for $t\geqslant 0$, so

$$p(t)(x(t)-x^*)\leqslant p^*(t)(x(t)-x^*) \qquad \text{for} \quad t\geqslant 0. \tag{3.14}$$

This yields in turn, for $t\geqslant 0$,

$$p(t)\,x(t)\leqslant p(t)\,x^* + p^*(t)(x(t)-x^*)\leqslant p(t)\,x^* + p^*(t)\,x(t). \tag{3.15}$$

Now, $p^*(t)\,x(t)\to 0$ as $t\to\infty$, since $p^*(t)=\delta^t q^*\to 0$ as $t\to\infty$, while $\|x(t)\|\leqslant B$ for $t\geqslant 0$. Also, by (3.8), $p(t)\,x^*\to 0$ as $t\to\infty$. Now, (3.13) follows from (3.15).

Since $\langle x(t), y(t), p(t)\rangle$ is a competitive program satisfying the transversality condition given by (3.13), so $\langle x(t), y(t)\rangle$ is optimal from $\tilde{y}$, by Proposition 2.1. ▮

Remark. The proof of Theorem 3.1 combines the ideas used in the earlier proofs (in Theorems 4.1 and 4.3) of Dasgupta and Mitra [3]. However, it uses weaker assumptions than either of those earlier results, and therefore includes both those results as special cases.

If we look at the proof of Theorem 3.1, we notice, in particular, that it does not use the convexity of the technology set or concavity of the welfare function. In fact, if we assume (A.1)–(A.4) and (A.6), and we also assume the condition

(E) *There exists an o.s.p.* (x^*, y^*), *with stationary price support*, q^*, *such that* x^* *is proportionately expansible*,

then Theorem 3.1 holds. In this sense, the requirements to prove Theorem 3.1 are similar to the requirements to prove Proposition 2.1. The *additional requirement* is condition (E), which can then be considered to be the "price paid" for replacing the transversality condition in Proposition 2.1 by the decentralizable condition in Theorem 3.1.

The converse of Theorem 3.1 is fairly well known. We state and prove it here formally for the sake of completeness.

THEOREM 3.2. *Suppose* $\langle x(t), y(t)\rangle$ *is an optimal program from* $\tilde{y}$ *in*

R^n_{++}. *Then, there is a sequence* $\langle p(t) \rangle$ *such that* $\langle x(t), y(t), p(t) \rangle$ *is competitive from* $\tilde{y}$, *and*

$$(q(t)-q^*)(y(t)-y^*) \leqslant 0 \quad \text{for} \quad t \geqslant 0. \tag{3.16}$$

Proof. By Proposition 2.2, there is a sequence $\langle p(t) \rangle$ such that $\langle x(t), y(t), p(t) \rangle$ is competitive from $\tilde{y}$, and for $t \geqslant 0$,

$$V(y(t)) - q(t)\, y(t) \geqslant V(y) - q(t)\, y \qquad \text{for all } y \text{ in } R^n_+. \tag{3.17}$$

Using $y = y^*$ in (3.17), we get for $t \geqslant 0$

$$V(y(t)) - V(y^*) \geqslant q(t)[y(t) - y^*]. \tag{3.18}$$

Using Proposition 2.4, $\langle x^*, y^*, p^*(t) \rangle$ is competitive from y^*, and $p^*(t)\, x^* = \delta^t q^* x^* \to 0$ as $t \to \infty$. Hence, by Proposition 2.3,

$$V(y^*) - q^* y^* \geqslant V(y) - q^* y \qquad \text{for all } y \text{ in } R^n_+. \tag{3.19}$$

Using $y = y(t)$ in (3.19), and transposing terms,

$$V(y^*) - V(y(t)) \geqslant -q^*[y(t) - y^*]. \tag{3.20}$$

Adding (3.18) and (3.20), we get (3.16). ∎

References

1. W. A. Brock and M. Majumdar, "On Characterizing Optimal Competitive Programs in Terms of Decentralizable Conditions," Cornell University Working Paper No. 333, 1985.
2. D. Cass and M. Majumdar, Efficient intertemporal allocation, consumption-value maximization, and capital-value transversality: A unified view, *in* "General Equilibrium, Growth and Trade" (J. R. Green and J. A. Scheinkman, Eds.), Academic Press, New York, 1979.
3. S. Dasgupta and T. Mitra, "Characterization of Intertemporal Optimality in Terms of Decentralizable Conditions: The Discounted Case," Cornell University Working Paper No. 346, 1985.
4. S. Dasgupta and T. Mitra, On price characterization of optimal plans in a multi-sector economy, mimeograph, Department of Economics, Cornell University, 1987.
5. D. Gale and W. R. S. Sutherland, Analysis of a one good model of economic development, *in* "Mathematics of the Decision Sciences" (G. B. Dantzig and A. F. Velnott, Jr., Eds.), Part 2, pp. 120–136, Amer. Math. Soc., Providence, RI, 1968.
6. M. A. Khan and T. Mitra, On the existence of a stationary optimal stock for a multi-sector economy: A primal approach, *J. Econ. Theory* **40** (1986), 319–328.
7. L. W. McKenzie, Optimal economic growth, turnpike theorems and comparative dynamics, *in* "Handbook of Mathematical Economics" Vol. III (K. J. Arrow and M. D. Intriligator, Eds.), North-Holland, New York, 1986.

8. B. Peleg, Efficiency prices for optimal consumption plans, III, *J. Math. Anal. Appl.* **32** (1970), 630–638.
9. B. Peleg, On competitive prices for optimal consumption plans, *SIAM J. Appl. Math.* **26** (1974), 239–253.
10. B. Peleg and H. E. Ryder, On optimal consumption plans in a multi-sector economy, *Rev. Econ. Stud.* **39** (1972), 159–169.
11. B. Peleg and H. E. Ryder, The modified golden rule of a multi-sector economy, *J. Math. Econ.* **1** (1974), 193–198.
12. B. Peleg and I. Zilcha, On competitive prices for optimal consumption plans, II, *SIAM J. Appl. Math.* **32** (1977), 127–131.
13. M. L. Weitzman, Duality theory for infinite horizon convex models, *Management Sci.* **19** (1973), 783–789.

5

Intertemporal Optimality in a Closed Linear Model of Production*

SWAPAN DASGUPTA

Department of Economics, Dalhousie University, Halifax, Nova Scotia, Canada B3H 3J5

AND

TAPAN MITRA

Department of Economics, Cornell University, Ithaca, New York 14853

The paper presents the main results on intertemporal optimality with discounting in a closed linear model of production, including the price characterization of optimal programs, the existence of a steady-state optimal program, and a turnpike property of optimal programs from arbitrary initial stocks. Some of these results are used to provide a characterization of the optimality of competitive programs in terms of a "decentralizable" condition. *Journal of Economic Literature* Classification Number: 111.

1. INTRODUCTION

The purpose of this paper is twofold. First, it tries to present the main results on intertemporal optimality of infinite horizon programs in a simple closed linear model of production. Second, it presents results on the characterization of the optimality of competitive programs in terms of a "decentralizable" condition (Theorems 1 and 2, Section 5).

Regarding the first aspect, we note that the results that we present are more or less well-known in the literature, particularly in the papers by McFadden [8] and Atsumi [1]. Our purpose is to present a systematic

* Research of the second author was supported by a National Science Foundation grant. Research on this paper was started while the first author was on sabbatic leave at Cornell University and Instituto Torcuato Di Tella, Buenos Aires. We are indebted to Mukul Majumdar for introducing us to the problems of intertemporal decentralization which this paper addresses. The present version has benefited from the detailed comments of two referees and an associate editor of the journal.

treatment in the context of a simple closed linear model, so that the main results are highlighted, while the additional complications which arise in more general models are avoided. These considerations dictate the choice of our production framework given by the simple linear model presented in Gale [5] and the preference framework given by iso-elastic utility functions as in Atsumi [1].

After discussing a von Neumann equilibrium in Section 3, we present the main results on optimal programs in Section 4. We first establish the existence of optimal programs using the existence criterion of Atsumi [1] and Brock and Gale [2]. We then present a "price characterization" of optimal programs: a feasible program is shown to be optimal if and only if it is competitive and it satisfies the transversality condition that the value of input stocks converges to zero. The necessity side of this result is contained in McFadden [8]. However, given our simple framework, we give an alternative proof, which exploits heavily the characteriation of *efficient* programs provided by Majumdar [9] and uses only the finite-dimensional separation theorem. Next, we obtain a result on the existence of a steady-state optimal program, that is, a program along which all outputs grow at a constant growth factor, and which is optimal in the usual sense. The line of argument used is essentially constructive, following Atsumi [1], but differentiability assumptions are not used on the welfare function. Finally, we use all of the above results to provide a "turnpike theorem" for optimal programs. That is, output produced along an optimal program is shown to be (a) of the same "composition" asymptotically as the steady-state optimal program and (b) growing at the same growth factor asymptotically as the steady-state optimal program.

Regarding the second aspect, we note that the problem here is to characterize the optimality of competitive programs in terms of a condition which can be verified under a decentralized system. A detailed discussion of and motivation for this problem can be found in Brock and Majumdar [3]. It suffices for our purpose to note that we want to replace the transversality condition in the price characterization results of Section 4 by a period-by-period condition, which involves for any given period the prices and quantities of the given competitive program and of the steady-state program for that period. We find that the condition used by Brock and Majumdar [3]—namely, that the scalar product of the price difference and the quantity difference be non-positive at each date—works for the closed linear model as well, but with one important qualification. The steady-state optimal stock is determined only up to positive scalar multiplication. If we fix the steady-state optimal stock (by suitable normalization), we must agree to compare it with competitive programs starting only from a certain set of initial stocks (the set being dependent on the choice of the steady-state optimal stock). The line of proof of the characterization result suggests that without this qualification, it is to be expected that the result (Theorem 1) would fail. We confirm this by studying a concrete example in the one-good version of our linear model (see Example 1, Section 5).

Proofs of our results are discussed only in Section 6.

2. Preliminaries

2a. *Notation*

Let R^n be an n-dimensional real space. For x, y in R^n, $x \geqslant y$ means $x_i \geqslant y_i$ for $i = 1, \ldots, n$; $x > y$ means $x \geqslant y$ and $x \neq y$; $x \gg y$ means $x_i > y_i$ for $i = 1, \ldots, n$. We denote the set $\{x \text{ in } R^n : x \geqslant 0\}$ by R^n_+, and the set $\{x \text{ in } R^n : x \gg 0\}$ by R^n_{++}.

We use the sum-norm on R^n; that is, the norm on R^n (denoted by $\|\cdot\|$) is defined by

$$\|x\| = \sum_{i=1}^{n} |x_i| \qquad \text{for all } x \text{ in } R^n.$$

The vector $(1, 1, \ldots, 1)$ in R^n is denoted by e. The ith unit vector in R^n is denoted by e^i, i.e., $e^i_j = 0$ for $i \neq j$ and $e^i_i = 1$; $i = 1, \ldots, n$, $j = 1, \ldots, n$. For any z in R^n, we denote $\min_i z_i$ by $m(z)$ and $\max_i z_i$ by $M(z)$.

Let A be an $n \times n$ real matrix. The generic element of A is denoted by a_{ij}, $i = 1, \ldots, n$ and $j = 1, \ldots, n$. The matrix, A, is called non-negative, and written as $A \geqslant 0$, if $a_{ij} \geqslant 0$ for $i = 1, \ldots, n$ and $j = 1, \ldots, n$. It is called positive, and written as $A > 0$, if $A \geqslant 0$ and A is not the null matrix. It is called strictly positive, and written as $A \gg 0$, if $a_{ij} > 0$ for $i = 1, \ldots, n$ and $j = 1, \ldots, n$.

2b. *The Model*

Consider an economy described by (Ω, w, δ), where Ω, a subset of $R^n_+ \times R^n_+$, is the technology set, $w: R^n_+ \to R$ is a welfare function, and δ is a discount factor, satisfying $0 < \delta < 1$.

We consider the technology to be of a simple polyhedral type, as specified in Gale [5],

$$\Omega = \{(x, y) \text{ in } R^n_+ \times R^n_+ : Ay \leqslant x\},$$

where A is an $n \times n$ real matrix, satisfying

(A.1) A is strictly positive; that is, $a_{ij} > 0$ for $i = 1, \ldots, n$ and $j = 1, \ldots, n$.

(A.2) A is productive; that is, there is y^0 in R^n_+ such that

$$Ay^0 \ll y^0.$$

Assumption (A.1) can be weakened somewhat using the theory of "primitive" matrices as discussed in Nikaido [10, pp. 108–114]. Assumption (A.2) is equivalent to the Hawkins–Simon condition (Nikaido [10, p. 90]).

Following Atsumi [1], we assume that the welfare function is of an iso-elastic type:

(A.3) $w(c)=[f(c)]^{1-\alpha}$ for c in R^n_+ where $0<\alpha<1$, and $f\colon R^n_+ \to R_+$ satisfies the following restrictions:

- (a) f is concave and continuous on R^n_+.
- (b) f is homogeneous of degree one.
- (c) $f(c')\geqslant f(c)$ when $c'\geqslant c$; $f(c')>f(c)$ if $c'>c$ and $f(c)>0$.
- (d) $f(c)>0$ for $c\gg 0$.

Remark 1. $w(c)\geqslant 0$ for c in R^n_+, since $f(c)\geqslant 0$ for c in R^n_+. Furthermore, from (A.3) (a) and (b) it follows that $f(0)=0$. Hence, $w(0)=0$.

A *program* from $\tilde{y}$ in R^n_+ is a sequence $\langle x(t), y(t)\rangle$, such that $y(0)=\tilde{y}$, $0\leqslant x(t)\leqslant y(t)$, and $(x(t), y(t+1))$ is in Ω for $t\geqslant 0$. Associated with a program $\langle x(t), y(t)\rangle$ is a *consumption sequence* $\langle c(t)\rangle$ given by

$$c(t)=y(t)-x(t) \qquad \text{for} \quad t\geqslant 0.$$

A program $\langle x(t), y(t)\rangle$ from $\tilde{y}$ is a *steady-state program* if $\tilde{y}>0$ and there is a real number, $g>0$, such that

$$y(t+1)=gy(t) \qquad \text{for} \quad t\geqslant 0.$$

A program $\langle x^*(t), y^*(t)\rangle$ from $\tilde{y}$ is an *optimal program* if

$$\sum_{t=0}^{\infty} \delta^t w(c^*(t))\geqslant \sum_{t=0}^{\infty} \delta^t w(c(t))$$

for all programs $\langle x(t), y(t)\rangle$ from $\tilde{y}$. A program $\langle x(t), y(t)\rangle$ from $\tilde{y}$ is a *steady-state optimal program* if it is a steady-state program and it is an optimal program from $\tilde{y}$. If there is a steady–state optimal program from $\tilde{y}$, we call $\tilde{y}$ a *steady-state optimal stock*.

A *competitive program* is a sequence $\langle x(t), y(t), p(t)\rangle$, such that

(1) $\langle x(t), y(t)\rangle$ is a program;

(2) $p(t)$ is in R^n_+ for $t\geqslant 0$;

(3) $$\delta^t w(c(t))-p(t)\,c(t)\geqslant \delta^t w(c)-p(t)\,c \qquad \text{for all } c \text{ in } R^n_+,\ t\geqslant 0 \qquad (2.1)$$

$$p(t+1)\,y(t+1)-p(t)\,x(t)\geqslant p(t+1)\,y-p(t)\,x \qquad \text{for all } (x, y) \text{ in } \Omega,\ t\geqslant 0. \qquad (2.2)$$

A competitive program $\langle x(t), y(t), p(t)\rangle$ is said to satisfy the *transversality condition* if

$$\lim_{t\to\infty} p(t)\,x(t)=0. \tag{2.3}$$

A program $\langle \bar{x}(t), \bar{y}(t)\rangle$ from $\tilde{y}$ is called *inefficient* if there is a program $\langle x(t), y(t)\rangle$ from $\tilde{y}$, such that $c(t)\geqslant \bar{c}(t)$ for all $t\geqslant 0$, and $c(t)>\bar{c}(t)$ for some $t\geqslant 0$. It is called *efficient* if it is not inefficient.

An obvious property of an efficient program $\langle x(t), y(t)\rangle$ is that for $t\geqslant 0$, $x(t)=Ay(t+1)$.

A useful characterization of the set of (feasible) programs and the set of efficient programs is provided in Theorem 4.1 of Majumdar [9]. We state it here for ready reference.

Result 1. *If* $\langle x(t), y(t)\rangle$ is a program from $\tilde{y}$ then $\sum_{t=0}^{\infty} A^t c(t)\leqslant \tilde{y}$. Conversely, if $c(t)$ is in R_+^n for $t\geqslant 0$, and $\tilde{y}$ is in R_+^n, and $\sum_{t=0}^{\infty} A^t c(t)\leqslant \tilde{y}$, then there is a program $\langle x'(t), y'(t)\rangle$ from $\tilde{y}$, with $c'(t)=c(t)$ for $t\geqslant 0$. A program $\langle x(t), y(t)\rangle$ from $\tilde{y}$ is efficient if and only if $\sum_{t=0}^{\infty} A^t c(t)=\tilde{y}$.

Note that the statement of Result 1 is somewhat stronger than that of Theorem 4.1 in Majumdar [9]. A careful reading of Majumdar's proof shows that the stronger statement is fully warranted. We state the result in this stronger form, because we find it most convenient to use it in this form.

Before we proceed to the substantive issues of the following sections, we must establish a preliminary result; viz., an optimal program from strictly positive initial stocks is efficient. This would be obvious if the welfare function was increasing in each component; this need not be the case "at the boundary" in our framework, and it would not be the case when f is of the Cobb–Douglas type.

Result 2. If $\langle \bar{x}(t), \bar{y}(t)\rangle$ is an optimal program from $\tilde{y}$ in R_{++}^n, then $\langle \bar{x}(t), \bar{y}(t)\rangle$ is an efficient program from $\tilde{y}$.

We shall also need the following well-known properties of "supporting prices" in the simple linear production model.

Result 3. Suppose (x^0, y^0) is in Ω, (p^0, q^0) is in $R_+^n\times R_+^n$, and $q^0y^0-p^0x^0\geqslant q^0y-p^0x$ for all (x, y) in Ω. Then the following hold:

(i) $q^0y^0-p^0x^0=0$.

(ii) $q^0-p^0A\leqslant 0$.

(iii) If $y^0\gg 0$ then $q^0=p^0A$.

(iv) If $p^0\gg 0$ then $Ay^0=x^0$.

3. A von Neumann Equilibrium

For any (x, y) in Ω with $x > 0$, let $\lambda(x, y) = \max\{\lambda : y \geqslant \lambda x\}$. It is known (see, for example, Karlin [7, p. 339]) that there is $(\hat{x}, \hat{y})$ in Ω (with $\hat{x} > 0$), $\hat{\lambda} > 0$, and a price vector $\hat{p} > 0$ such that

(i) $\hat{\lambda} = \lambda(\hat{x}, \hat{y})$, $\hat{y} = \hat{\lambda}\hat{x}$

(ii) $\hat{\lambda} \geqslant \lambda(x, y)$ for all (x, y) in Ω with $x > 0$

(iii) $\hat{p}y \leqslant \hat{\lambda}\hat{p}x$ for all (x, y) in Ω.

We refer to $\hat{x}$ as a vector of *von Neumann stocks*, $\hat{\lambda}$ as the *von Neumann growth factor*, and $\hat{p}$ as a *von Neumann price*. We refer to $(\hat{x}, \hat{y}, \hat{\lambda}, \hat{p})$ as a *von Neumann equilibrium.*

We now relate $\hat{\lambda}$ to the Frobenius eigenvalue of A and $\hat{y}$, $\hat{p}$ to the Frobenius eigenvectors of A.

Since $\hat{x} > 0$, $\hat{\lambda} > 0$, and $\hat{y} = \hat{\lambda}\hat{x}$, so $\hat{y} > 0$. Since $(\hat{x}, \hat{y})$ is in Ω, so $\hat{x} \geqslant A\hat{y}$; and since $\hat{y} > 0$, while $A \gg 0$, so we have $\hat{x} \gg 0$. Since $\hat{y} = \hat{\lambda}\hat{x}$, so $\hat{y} \gg 0$. From (i) and (iii), we have

$$\hat{p}\hat{y} - \hat{\lambda}\hat{p}\hat{x} = 0 \geqslant \hat{p}y - \hat{\lambda}\hat{p}x \qquad \text{for all } (x, y) \text{ in } \Omega. \tag{3.1}$$

Since $\hat{y} \gg 0$, therefore, by Result 3(iii) we have

$$\hat{p} = \hat{\lambda}\hat{p}A, \qquad \text{that is (since } \hat{\lambda} > 0\text{)},$$

$$(1/\hat{\lambda})\,\hat{p} = \hat{p}A. \tag{3.2}$$

Thus, $(1/\hat{\lambda})$ is an eigenvalue of A, and $\hat{p}$ a corresponding eigenvector. Since $\hat{p} > 0$, so $\hat{p}$ is the (left-hand) Frebenius eigenvector of A, and $(1/\hat{\lambda})$ is the (simple) Frobenius eigenvalue of A (Gantmacher [6, p. 63]).

Since $\hat{p} > 0$, and $A \gg 0$, so (3.2) implies that $\hat{p} \gg 0$. Therefore, by Result 3(iv) we get $\hat{x} = A\hat{y}$, so that we get

$$(1/\hat{\lambda})\,\hat{y} = A\hat{y}. \tag{3.3}$$

Since $\hat{y} > 0$, so $\hat{y}$ is the right-hand Frobenius eigenvector of A (Gantmacher [6, p. 63]).

In what follows, we normalize $\hat{y}$, *so that* $\|\hat{y}\| = 1$. *We then normalize* $\hat{p}$, *so that* $\hat{p}\hat{y} = 1$.

We note that since A is productive, [see (A.2) above], there is y^0 such that (Ay^0, y^0) is in Ω, with $y^0 > 0$, and $Ay^0 \ll y^0$. Thus, $\lambda(Ay^0, y^0) > 1$, so that $\hat{\lambda} \geqslant \lambda(Ay^0, y^0) > 1$.

4. Optimality

This section summarizes the main results on optimal growth with discounting in the closed linear model of production.

We start by providing a sufficient condition for the *existence of an optimal program* in our framework. This condition is then maintained for the rest of the discussion. The condition is

(A.4) $\delta\hat{\lambda}^{(1-\alpha)} < 1.$

Discussions of the interpretation of (A.4) are available in Brock and Gale [2] and McFadden [8]. Our existence result can be stated as follows.

Proposition 1. *Given an initial stock $\tilde{y}$ in R^n_+, if $\langle x(t), y(t)\rangle$ is any program from $\tilde{y}$, then*

$$\sum_{t=0}^{\infty} \delta^t w(c(t)) < \infty.$$

Furthermore, there is an optimal program $\langle x^*(t), y^*(t)\rangle$ from $\tilde{y}$.

In view of the existence result above, we can define a *value function*, $V\colon R^n_+ \to R$, by

$$V(y) = \sum_{t=0}^{\infty} \delta^t w(c^*(t)) \qquad \text{for } \ y \text{ in } R^n_+,$$

where $\langle x^*(t), y^*(t)\rangle$ is an optimal program from y.

If $\langle x^*(t), y^*(t)\rangle$ is an optimal program from $\tilde{y}$ in R^n_+, then the "principle of optimality" states that for $N \geqslant 0$,

$$V(y^*(0)) = \sum_{t=0}^{N} \delta^t w(c^*(t)) + \delta^{N+1} V(y^*(N+1)).$$

The proof of this principle is too well-known to be reproduced here in detail.

We now note the main results on the price characterization of optimal programs. A competitive program is optimal if it satisfies the transversality condition that the value of input stocks converges to zero (at least for a subsequence of periods).

Proposition 2. *Suppose $\langle \bar{x}(t), \bar{y}(t), \bar{p}(t)\rangle$ is a competitive program from $\tilde{y}$ in R^n_+, and*

$$\liminf_{t\to\infty} \bar{p}(t)\,\bar{x}(t) = 0$$

then $\langle \bar{x}(t), \bar{y}(t)\rangle$ is an optimal program from $\tilde{y}$.

Conversely, an optimal program is competitive and satisfies the transversality condition that the value of input stocks coverges to zero.

PROPOSITION 3. *Suppose* $\langle \bar{x}(t), \bar{y}(t) \rangle$ *is an optimal program from* $\tilde{y}$ *in* R^n_{++}. *Then, there is a sequence* $\langle \bar{p}(t) \rangle$ *such that* $\langle \bar{x}(t), \bar{y}(t), \bar{p}(t) \rangle$ *is a competitive program,* $\bar{p}(t) > 0$ *for* $t \geqslant 0$, *and*

$$\lim_{t \to \infty} \bar{p}(t)\, \bar{x}(t) = 0.$$

Remark 2. Our proof of Proposition 3 exploits the simple structure of the production model. It essentially uses the method of Weitzman [11], of obtaining a "price support" for the value function in the *initial* period, but avoids the induction argument for obtaining the "supporting prices" in the subsequent periods for the technology and the value function, by exploiting the structure of the simple polyhedral model. For a more general linear model, a similar result can be proved, following the method of McFadden [8, Theorem 5, pp. 47–48].

It is worth noting that the "prices" $\langle \bar{p}(t) \rangle$ in Proposition 3 also support the value function at $\bar{y}(t)$.

PROPOSITION 4. *Suppose* $\langle \bar{x}(t), \bar{y}(t), \bar{p}(t) \rangle$ *is a competitive program from* $\tilde{y}$ *in* R^n_+ *and* $\lim_{t \to \infty} \bar{p}(t)\, \bar{x}(t) = 0$. *Then*

$$\delta^t V(\bar{y})) - \bar{p}(t)\, \bar{y}(t) \geqslant \delta^t V(y) - \bar{p}(t)\, y \qquad \textit{for all } y \textit{ in } R^n_+,\ t \geqslant 0.$$

COROLLARY. *Suppose* $\langle \bar{x}(t), \bar{y}(t) \rangle$ *is an optimal program from* $\tilde{y}$ *in* R^n_{++}. *Then, there is a sequence* $\langle \bar{p}(t) \rangle$, *such that* $\langle \bar{x}(t), \bar{y}(t), \bar{p}(t) \rangle$ *is a competitive program, and*

(i) $\lim_{t \to \infty} \bar{p}(t)\, \bar{x}(t) = 0$

(ii) $\delta^t V(\bar{y}(t)) - \bar{p}(t)\, \bar{y}(t) \geqslant \delta^t V(y) - \bar{p}(t)\, y$ *for all* y *in* R^n_+, $t \geqslant 0$.

Next, we turn to a result on the *existence of a steady-state optimal program*. We show the existence of a stock y^*, such that the program from y^* along which stocks grow at the growth factor

$$g = (\delta \hat{\lambda})^{(1/\alpha)}$$

[that is, along which $y^*(t) = g^t y^*$ for $t \geqslant 0$] is optimal among all programs from y^*.

PROPOSITION 5. *There exists* $y^* \gg 0$, *such that* $y^*(t) = g^t y^*$, $x^*(t) = Ay^*(t+1)$ *for* $t \geqslant 0$ *defines a steady-state optimal program from* y^*, *where*

$$g = (\delta \hat{\lambda})^{(1/\alpha)}.$$

Remark 3. If we define for $\lambda > 0$, $< x(t), y(t), p(t) >$ by $[x(t), y(t), p(t)] = [\lambda x^*(t), \lambda y^*(t), \lambda^{-\alpha} p^*(t)]$ for $t \geqslant 0$, then it follows that $\langle x(t), y(t), p(t) \rangle$ is competitive, and $\lim_{t \to \infty} p(t)\, x(t) = \lim_{t \to \infty} \lambda^{1-\alpha} p^*(t)\, x^*(t) = 0$. Therefore, by Proposition 2, $\langle x(t), y(t) \rangle$ is optimal. Clearly it is also a steady-state program. Hence $\langle x(t), y(t) \rangle$ is a steady-state optimal program. [More generally, it is easily seen by a similar argument that if $\langle \bar{x}(t), \bar{y}(t) \rangle$ is an optimal program from $\bar{y}$, then for any $\lambda > 0$, $\langle \lambda \bar{x}(t), \lambda \bar{y}(t) \rangle$ is an optimal program from $\lambda \tilde{y}$.]

The significance of the steady-state optimal program (of Proposition 5) for optimal programs in general is best conveyed through a *turnpike property of optimal programs*. To establish this, we need to strengthen our assumptions on the welfare function. If c and c' are in R^n then by "c is *proportional to* c'," we mean that there is a real number $\mu \neq 0$ such that $c = \mu c'$; otherwise, *c is not proportional to c'*. We now assume, in addition to (A.1)–(A.4), the following property.

(A.5) If c, c' are in R^n_+, c is not proportional to c' and $f(c) > 0 < f(c')$, then, for $0 < \theta < 1$, $f(\theta c + (1-\theta)\, c') > \theta f(c) + (1-\theta) f(c')$.

PROPOSITION 6. *Suppose* $\langle \bar{x}(t), \bar{y}(t) \rangle$ *is an optimal program from* $\tilde{y} \gg 0$. *Then, there is a positive real number* μ, *such that* (i) $(x'(t), y'(t)) = g^t \mu(gAy^*, y^*)$ *for* $t \geqslant 0$ *defines a steady-state optimal program from* μy^*; (ii) $\lim_{t \to \infty} [\bar{y}(t)/g^t] = \mu y^*$ [*where* $y^* \gg 0$ *is the steady-state optimal stock and* g *is the steady-state growth factor obtained in Proposition* 5].

The above turnpike result implies two things about the behavior of stocks along an optimal program $\langle \bar{x}(t), \bar{y}(t) \rangle$. First, the *composition* of the stocks, $[y(t)/\| y(t) \|]$, converges to the *composition* of the steady-state optimal stock, $[y^*/\| y^* \|]$. Second, the *growth factor* of the stocks $[\| y(t+1) \| / \| y(t) \|]$ converges to the *growth factor* of the steady-state optimal program, g.

Since for the optimal program $\langle \bar{x}(t), \bar{y}(t) \rangle$ we will have $\bar{x}(t) = A\bar{y}(t+1)$, so it follows trivially from Proposition 6 that

$$[\bar{x}(t)/g^t] \to \mu x^*(0) \text{ as } t \to \infty \qquad (\text{where } x^*(0) = gAy^*)$$

and

$$[\bar{c}(t)/g^t] \to \mu c^*(0) \text{ as } t \to \infty,$$

where $c^*(0)$ is defined as $[y^* - x^*(0)]$.

5. DECENTRALIZATION

In this section, we show that optimality of competitive programs can be characterized in terms of the simple decentralizable rule of Brock–Majumdar [3]. However, there is one important qualification. There is a potential choice of the steady–state optimal program (i.e., a choice regarding the appropriate positive scalar multiple) in terms of which the decentralizable rule [(5.1) below] is to be stated. The necessity side of the characterization result (Theorem 2) and Remark 5 show that, for an optimal program from any $\tilde{y}$ in R^n_{++} the decentralizable rule (5.1) is satisfied for *every* steady-state optimal program. Conversely, (as Theorem 1 shows) if $\langle \bar{x}(t), \bar{y}(t), \bar{p}(t) \rangle$ is a competitive program and the decentralizable rule (5.1) is satisfied with respect to *every* steady-state optimal program $\langle x^*(t), y^*(t) \rangle$ then $\langle \bar{x}(t), \bar{y}(t) \rangle$ is an optimal program. However, this requires verification of the rule, in principle, in infinitely many cases (corresponding to the infinitely many steady-state optimal programs). It is obviously of interest to state the rule in terms of *one particular* steady-state optimal program. A plausible conjecture may be that the rule could be stated in terms of any one of the steady-state optimal programs; in other words, if the rule is satisfied for some steady-state optimal program then (given competitiveness) the program is optimal. If this were true then this would be the simplest possible way to state the rule. In the course of the proof of Theorem 1, i.e., in the course of showing that, if a competitive program is not optimal then there is a steady-state program for which (5.1) is violated, it becomes clear, however, that the steady-state program cannot be arbitrarily chosen. This is confirmed by the example following Theorem 2. It is necessary, therefore, to choose an appropriate steady-state optimal program (given the competitive program), in terms of which, if the decentralizable rule is stated, it signals optimality. Equivalently, given any particular steady-state optimal program, one can define a set of initial stocks (the set Y below), for which the decentralizable rule (for competitive programs from such initial stocks), in terms of the given steady-state program, does signal optimality. We follow this latter approach in Theorem 1.

Denote $\hat{p}\hat{y}/\hat{p}c^*$ by E. Since $c^* > 0$ and $\hat{p} \gg 0$, E is well defined, and since $\hat{y} \gg 0$, $E > 0$. Next, define $\eta = [m(\hat{y})/E \, \| c^* \|]$, and denote

$$\theta = [\eta/(\eta + 1)]^{[1/\alpha(1-\alpha)]}.$$

Since $\hat{y} \gg 0$, we have, $\eta > 0$ and $0 < \theta < 1$. Now we define a set of initial stocks for which the exercise will be carried out:

$$Y = \{ y \text{ in } R^n_+ : \hat{p}y = \theta \hat{p} c^* \}.$$

THEOREM 1. *Suppose* $\langle \bar{x}(t), \bar{y}(t), \bar{p}(t) \rangle$ *is a competitive program from* $\tilde{y}$ *in* Y. *Suppose, for* $t \geqslant 0$,

$$[\bar{p}(t) - p^*(t)][\bar{y}(t) - y^*(t)] \leqslant 0 \tag{5.1}$$

then $\langle \bar{x}(t), \bar{y}(t) \rangle$ *is optimal from* $\tilde{y}$.

Remark 4. Inspecting the rule (5.1), it appears that besides information regarding the "current" commodity prices measured in terms of current utility units $[(1/\delta^t)\,p(t)]$, the output at time t along the competitive program, $[y(t)]$, and $\hat{p}$ and y^*, it is also necessary to know the value of t [for computing $y^*(t)$] in order to be able to verify (5.1). This is an awkward requirement, since the period which is regarded as the origin of measurement of time should not be of any significance and would be an odd information requirement from the point of view of agents at time t. It would seem to be desirable that the rule be in a form where, the information required for its verification is the current prices and current quantities along the competitive program and the information regarding the normalized steady-state compositions, viz., $\hat{p}$, c^* and y^*, and α.

If $\langle p(t)\rangle$ is a sequence of present value prices, let $\langle q(t)\rangle$, defined by $q(t)\equiv(1/\delta^t)\,p(t)$, be the corresponding sequence of current prices. Suppose that we have the current price and quantity information along a competitive program $\langle \bar{x}(t),\ \bar{y}(t),\ \bar{p}(t)\rangle$, viz., $\bar{x}(t)$, $\bar{y}(t)$, and $\bar{q}(t)$. We wish to select $\tilde{q}(t)$ and $\tilde{y}(t)$ for verification of the rule (5.1), rewritten as $[\bar{q}(t)-\tilde{q}(t)][\bar{y}(t)-\tilde{y}(t)]\leqslant 0$ for $t\geqslant 0$. We want this selection to correspond to an optimal steady-state program, which, if used in the verification of (5.1) (in the form just stated), does signal optimality for the given competitive program. Define $\gamma(t)=\frac{1}{2}\bar{q}(t)\,\hat{y}$. Now $\bar{q}(t)\,\hat{y}=(1/\delta^t)\,\bar{p}(t)\,\hat{y}>0$, since $\bar{p}(t)>0$; hence, $\gamma(t)>0$. Denote $(1/\gamma(t)^{1/\alpha})$ by $\beta(t)$ and define $\tilde{c}(t)=\beta(t)\,c^*$, $\tilde{y}(t)=\beta(t)\,y^*$, $\tilde{x}(t)=\beta(t)\,x^*$ (where $x^*=gAy^*$), $\tilde{q}(t)=\gamma(t)\,\hat{p}$. Then it can be shown that $\langle \tilde{x}(t),\ \tilde{y}(t)\rangle$ is an optimal steady-state program with associated present value prices $\tilde{p}(t)=\delta^t\tilde{q}(t)$. Furthermore, it can be checked (essentially by following the method used to prove Theorem 1) that if

$$[\bar{q}(t)-\tilde{q}(t)][\bar{y}(t)-\tilde{y}(t)]\leqslant 0$$

then $\langle \bar{x}(t),\ \bar{y}(t)\rangle$ is an optimal program from $\hat{y}$.

The above approach, namely that of choosing the appropriate steady-state program, given the competitive program, is equivalent to the approach of Theorem 1. It should be emphasized that, in defining the steady–state values in period t, the value of t was not used. Finally, it may also be remarked that, denoting $v(t)=[\bar{p}(t)-p^*(t)][\bar{y}(t)-y^*(t)]$ for $t\geqslant 0$, it can be checked that $v(t+1)-v(t)\geqslant 0$ for all $t\geqslant 0$ along a competitive program. Hence, the proof of Theorem 1 actually shows that, if the given competitive program is not optimal, then $v(t)>0$ for all but finitely many periods.

A converse of the result of Theorem 1 can now be noted.

THEOREM 2. *Suppose $\langle \bar{x}(t),\ \bar{y}(t)\rangle$ is an optimal program from $\hat{y}$ in Y*

and $\tilde{y} \gg 0$. *Then, there is a sequence* $\langle \bar{p}(t) \rangle$ *such that* $\langle \bar{x}(t), \bar{y}(t), \bar{p}(t) \rangle$ *is a competitive program, and for* $t \geqslant 0$

$$[\bar{p}(t) - p^*(t)][\bar{y}(t) - y^*(t)] \leqslant 0.$$

Remark 5. We note that the fact that $\tilde{y}$ is in Y is of no significance in Theorem 2; the result is true without this restriction. However, it is extremely significant in Theorem 1 where the result is not necessarily true without this restriction. We justify this last statement now with an example.

EXAMPLE 1. Let us consider a one-good example, in which $\Omega = \{(x, y)$ in $R_+ \times R_+ :\ x \geqslant ay\}$. Here $0 < a < 1$ ensures that (A.1) and (A.2) are satisfied. The welfare function is $w(c) = c^{1-\alpha}$ (where $0 < \alpha < 1$) for c in R_+; this ensures that (A.3) is satisfied. The discount factor is $0 < \delta < 1$.

It is easy to check that $\hat{y} = 1$, $\hat{x} = a$, $\hat{\lambda} = (1/a)$, $\hat{p} = 1$ is a von Neumann equilibrium, satisfying our stipulated normalizeation of $\hat{y}$ and $\hat{p}$. Assuming $\delta < a^{1-\alpha}$ ensures that (A.4) is satisfied. For the sake of concreteness, choose $a = (\frac{1}{4})$, $\alpha = \frac{1}{2}$, $\delta = \frac{1}{4}$, so that $\hat{\lambda} = 4 = (1/\hat{x})$, and $a^{1-\alpha} = (\frac{1}{4})^{1/2} = \frac{1}{2} > \frac{1}{4} = \delta$.

Define $g = (\delta\hat{\lambda})^{1/\alpha} = 1$, $c^* = (1-\alpha)^{1/\alpha} = \frac{1}{4}$, $y^* = c^*(1-ga)^{-1} = \frac{1}{3}$. Note then that $\langle x^*(t), y^*(t) \rangle$ defined by $y^*(t) = \frac{1}{3}$, $x^*(t) = \frac{1}{12}$ for $t \geqslant 0$ is a steady-state program from $y^* = \frac{1}{3}$. Also, $\langle x^*(t), y^*(t), p^*(t) \rangle$ is a competitive program, where $p^*(t) = \frac{1}{4^t}$ for $t \geqslant 0$, and $p^*(t)\, y^*(t) \to 0$ as $t \to \infty$. So $\langle x^*(t), y^*(t) \rangle$ is an optimal steady-state program.

Next, let $\tilde{y} > 0$, and define $\langle \bar{x}(t), \bar{y}(t) \rangle$ by $\bar{y}(t) = \tilde{y}(\hat{\lambda}^t + g^t)/2 = \tilde{y}(4^t + 1)/2$ for $t \geqslant 0$, $\bar{x}(t) = \bar{y}(t+1)/\hat{\lambda} = \bar{y}(t+1)/4$ for $t \geqslant 0$. Then, $\bar{c}(t) = \bar{y}(t) - [\bar{y}(t+1)/\hat{\lambda}] = [\tilde{y}(\hat{\lambda}^t + g^t)/2] - [\tilde{y}(\hat{\lambda}^{t+1} + g^{t+1})/2\hat{\lambda}] = (\tilde{y}/2\hat{\lambda})[\hat{\lambda}^{t+1} + \hat{\lambda}g^t - \hat{\lambda}^{t+1} - g^{t+1}] = (\tilde{y}/2\hat{\lambda})(\hat{\lambda} - g)\, g^t = \frac{3}{8}\tilde{y}$ for $t \geqslant 0$. Define $\langle \bar{p}(t) \rangle$ as follows: $\bar{p}(t) = \delta^t(1-\alpha)[\bar{c}(t)]^{-\alpha} = \frac{1}{2}(8/3\tilde{y})^{1/2}/4^t$ for $t \geqslant 0$. One can check that $\langle \bar{x}(t), \bar{y}(t), \bar{p}(t) \rangle$ is a competitive program. Clearly $\bar{p}(t)\ \bar{y}(t) \to (8/3\tilde{y})^{1/2}\ (\tilde{y}/4)$ as $t \to \infty$, and [since $\langle \bar{p}(t) \rangle$ are the unique competitive prices] $\langle \bar{x}(t), \bar{y}(t) \rangle$ is not an optimal program from $\tilde{y}$.

Now, if we choose $\tilde{y} = 1$, then $\bar{p}(t) = (\frac{2}{3})^{1/2}/4^t < 1/4^t = p^*(t)$ for $t \geqslant 0$. Also, $\bar{y}(t) = (4^t + 1)/2 > \frac{1}{3} = y^*(t)$ for $t \geqslant 0$. So, for $t \geqslant 0$,

$$[\bar{p}(t) - p^*(t)][\bar{y}(t) - y^*(t)] \leqslant 0$$

but, as we have already noted, $\langle \bar{x}(t), \bar{y}(t) \rangle$ is not optimal from $\tilde{y} = 1$. Thus, the above rule fails to signal the non-optimality of the competitive program, $\langle \bar{x}(t), \bar{y}(t), \bar{p}(t) \rangle$.

The problem is, as we have mentioned earlier in the discussion, that the initial stock $\tilde{y}$ was not chosen carefully enough, given the comparison steady-state program. In this example, $E = 4$, $\eta = 1$, $\theta = (\frac{1}{2})^4 = \frac{1}{16}$, and $Y = \{y : y = \frac{1}{64}\}$. Now, if $\tilde{y}$ is chosen in Y, that is $\tilde{y} = \frac{1}{64}$, then

$\bar{p}(t)=(\frac{128}{3})^{1/2}/4^t$ for $t\geqslant 0$, so that $\bar{p}(t)>p^*(t)$ for $t\geqslant 0$. Also, $\bar{y}(t)=(4^t+1)/128>\frac{1}{3}=y^*(t)$ for $t\geqslant 3$. So, for $t\geqslant 3$,

$$[\bar{p}(t)-p^*(t)][\bar{y}(t)-y^*(t)]>0$$

and the above rule signals the non-optimality of the competitive program $\langle\bar{x}(t), \bar{y}(t), \bar{p}(t)\rangle$.

We note that the definition of the initial stocks (the set Y), given the steady state, does not completely characterize the set of initial stocks for which the decentralizable rule works (given the steady state). The proof of Theorem 1 makes it clear that the essential restriction on the initial stock is that all competitive programs starting from it must satisfy the following inequality:

$$\bar{p}(0)\,\hat{y}>\hat{p}\hat{y}. \tag{5.2}$$

Initial stocks which are "slightly" larger than those included in the definition of Y may satisfy this. If $\tilde{y}=1$, $\bar{p}(0)\,\hat{y}=(\frac{2}{3})^{1/2}<1=\hat{p}\hat{y}$, so that inequality (5.2) is violated. If $\tilde{y}$ is chosen to be $\frac{2}{3}$ then $\bar{p}(0)\,\hat{y}=1=\hat{p}\hat{y}$, this being the borderline case where (5.2) is still violated. Here $\bar{p}(t)=1/4^t=p^*(t)$. Hence $[\bar{p}(t)-p^*(t)][\bar{y}(t)-y^*(t)]=0$ for $t\geqslant 0$ so that the rule (5.1) still fails to signal non-optimality of $\langle\bar{x}(t), \bar{y}(t)\rangle$. If $\tilde{y}<\frac{2}{3}$, then (5.2) is satisfied and $[\bar{p}(t)-p^*(t)]>0$ for $t\geqslant 0$. It is clear from the definition of $\bar{y}(t)$ that so long as $\tilde{y}>0$, $\bar{y}(t)-y^*(t)>0$ for t sufficiently large and hence (5.1) is violated for t sufficiently large; that is, (5.1) does signal non-optimality. The above discussion illustrates that the critical consideration in the the choice of $\tilde{y}$ (that is, in the definition of Y) is that *all* competitive programs emanating from any $\tilde{y}$ in Y must satisfy (5.2).

6. Proofs

In this section, we provide the proofs of the main results in Sections 2–5 For the detailed proofs of all the results, the reader is referred to the working paper by Dasgupta and Mitra [4].

Proof of Result 2

Let $\langle\bar{x}(t), \bar{y}(t)\rangle$ be an optimal program from $\tilde{y}\gg 0$. Then, there is some time period, s, for which $w(\bar{c}(s))>0$. We claim that if $s\geqslant 1$, then $w(\bar{c}(s-1))>0$ also. If not, then $w(\bar{c}(s-1))=0$. Choose $0<\lambda<1$, with λ sufficiently close to 1, so that

$$[w(A\bar{c}(s))/(1-\lambda)^{\alpha}]\geqslant\delta w(\bar{c}(s)). \tag{6.1}$$

Note that $\bar{c}(s) > 0$, so $A\bar{c}(s) \gg 0$, and so $w(A\bar{c}(s)) > 0$, so that by suitable choice of λ, the inequality (6.1) can be satisfied.

Consider a sequence $\langle x'(t), y'(t)\rangle$ defined by $[x'(t), y'(t)] = [\bar{x}(t), \bar{y}(t)]$ for $t \neq s, s-1$; $y'(s-1) = \bar{y}(s-1)$, $y'(s) = [\lambda\bar{c}(s) + \bar{x}(s)]$; $x'(s-1) = Ay'(s)$, $x'(s) = \bar{x}(s)$. Note that $y'(s) \geqslant x'(s)$, and $x'(s-1) = Ay'(s) \leqslant A\bar{y}(s) \leqslant \bar{x}(s-1) \leqslant \bar{y}(s-1) = y'(s-1)$. Hence, $\langle x'(t), y'(t)\rangle$ is a program from y. Also, $c'(t) = \bar{c}(t)$ for $t \neq s, s-1$; $c'(s) = \lambda\bar{c}(s)$, and $c'(s-1) = y'(s-1) - x'(s-1) = \bar{y}(s-1) - Ay'(s) = \bar{y}(s-1) - A\bar{x}(s) - \lambda A\bar{c}(s) - (1-\lambda)A\bar{c}(s) + (1-\lambda)A\bar{c}(s) = \bar{y}(s-1) - A[\bar{x}(s) + \bar{c}(s)] + (1-\lambda)A\bar{c}(s) = \bar{y}(s-1) - A\bar{y}(s) + (1-\lambda)A\bar{c}(s) \geqslant \bar{c}(s-1) + (1-\lambda)A\bar{c}(s) \geqslant (1-\lambda)A\bar{c}(s)$.

Then, $w(c'(s-1)) + \delta w(c'(s)) \geqslant w((1-\lambda)A\bar{c}(s)) + \delta w(\lambda\bar{c}(s)) = (1-\lambda)^{1-\alpha} w(A\bar{c}(s)) + \delta\lambda^{1-\alpha} w(\bar{c}(s)) > (1-\lambda)[w(A\bar{c}(s))/(1-\lambda)^{\alpha}] + \delta\lambda\, w(\bar{c}(s))$ [since $\lambda^{1-\alpha} > \lambda$ and $w(\bar{c}(s)) > 0$] $\geqslant (1-\lambda)\delta w(\bar{c}(s)) + \delta\lambda w(\bar{c}(s))$ [using (6.1)] $= \delta w(\bar{c}(s)) = w(\bar{c}(s-1)) + \delta w(\bar{c}(s))$. This shows that [since $c'(t) = \bar{c}(t)$ for $t \neq s, s-1$] $\langle \bar{x}(t), \bar{y}(t)\rangle$ is not optimal, a contradiction which proves our claim.

In view of this, repeating the above argument for a finite number of periods, we can conclude that $w(\bar{c}(0)) > 0$.

Now, we claim that $\langle \bar{x}(t), \bar{y}(t)\rangle$ is efficient. Otherwise, by Result 1,

$$\sum_{t=0}^{\infty} A^t\bar{c}(t) < \tilde{y}.$$

Defining $c(0) = \bar{c}(0) + [\tilde{y} - \sum_{t=0}^{\infty} A^t\bar{c}(t)] > \bar{c}(0)$, and $c(t) = \bar{c}(t)$ for $t \geqslant 1$, we note that

$$\sum_{t=0}^{\infty} A^t c(t) = \tilde{y}.$$

So, by Result 1, there is a program $\langle x'(t), y'(t)\rangle$ from $\tilde{y}$ with $c'(t) = c(t)$ for $t \geqslant 0$. Now, $w(c'(t)) = w(c(t)) = w(\bar{c}(t))$ for $t \geqslant 1$, and $w(c'(0)) = w(c(0)) > w(\bar{c}(0))$, using (A.3) and the facts that $c(0) > \bar{c}(0)$ and $w(\bar{c}(0)) > 0$ [and so $f(\bar{c}(0)) > 0$]. This shows that $\langle \bar{x}(t), \bar{y}(t)\rangle$ is not optimal from $\tilde{y}$, a contradiction. Hence $\langle \bar{x}(t), \bar{y}(t)\rangle$ is efficient. ∎

Proof of Proposition 1

Given an initial stock, $\tilde{y}$ in R^n_+, define $B = [1/m(\hat{p})]$, $\hat{B} = [w(B\hat{p}\tilde{y}e)/(1 - \delta\hat{\lambda}^{1-\alpha})]$ and a sequence $\langle k(t)\rangle$ by $k(t) = \hat{\lambda}^t(B\hat{p}\tilde{y})\,e$ for $t \geqslant 0$. Note that $\sum_{t=0}^{\infty} \delta^t w(k(t))$ is a convergent geometric series [given (A.4)] and clearly

$$\sum_{t=0}^{\infty} \delta^t w(k(t)) = \hat{B}.$$

Next, let $\langle x(t), y(t)\rangle$ be a program from $\tilde{y}$. Then, for $t \geqslant 0$,

$$\hat{p}y(t+1) = \hat{\lambda}\hat{p}Ay(t+1) \leqslant \hat{\lambda}\hat{p}x(t) \leqslant \hat{\lambda}\hat{p}y(t)$$

so that $\hat{p}y(t) \leqslant \hat{\lambda}^t\hat{p}\tilde{y}$ for all $t \geqslant 0$. Thus, for $i = 1, ..., n$, $y_i(t) \leqslant \hat{\lambda}^t B\hat{p}\tilde{y}$ for $t \geqslant 0$, and so $y(t) \leqslant \hat{\lambda}^t(B\hat{p}\tilde{y})\, e \equiv k(t)$ for $t \geqslant 0$. Since $c(t) \leqslant y(t)$, so $c(t) \leqslant k(t)$ and $w(c(t)) \leqslant w(k(t))$ for $t \geqslant 0$.

It is, of course, clear that $\sum_{t=0}^{T} \delta^t w(c(t))$ is bounded above by $\hat{B}$ and is monotonically non-decreasing in T, so it converges, and $\sum_{t=0}^{\infty} \delta^t w(c(t)) \leqslant \hat{B}$.

Let $\langle x(t), y(t)\rangle$ be any program from $\tilde{y}$. Then

$$\sum_{t=0}^{\infty} \delta^t[w(k(t)) - w(c(t))] \leqslant \sum_{t=0}^{\infty} \delta^t w(k(t)) = \hat{B}.$$

So, by using Brock–Gale [2, Lemma 2, p. 236], there is a program $\langle x^*(t), y^*(t)\rangle$ from $\tilde{y}$, such that

$$\sum_{t=0}^{\infty} \delta^t[w(k(t)) - w(c^*(t))] \leqslant \sum_{t=0}^{\infty} \delta^t[w(k(t)) - w(c(t))]$$

for every program $\langle x(t), y(t)\rangle$ from $\tilde{y}$. This means

$$\sum_{t=0}^{\infty} \delta^t w(c^*(t)) \geqslant \sum_{t=0}^{\infty} \delta^t w(c(t))$$

for every program $\langle x(t), y(t)\rangle$ from $\tilde{y}$. Hence $\langle x^*(t), y^*(t)\rangle$ is an optimal program from $\tilde{y}$. ∎

Proof of Proposition 3

Define the sets G and H as follows:

$$G = \left\{ (a, b) \text{ in } R \times R^n\colon a \leqslant \sum_{t=0}^{\infty} \delta^t w(c(t)) - \sum_{t=0}^{\infty} \delta^t w(\bar{c}(t)), \right.$$
$$b \leqslant \sum_{t=0}^{\infty} A^t \bar{c}(t) - \sum_{t=0}^{\infty} A^t c(t),$$
$$\left. \text{for some program } \langle x(t), y(t)\rangle \right\}. \tag{6.2}$$

It is worth emphasizing that the program referred to in the definition of G, $\langle x(t), y(t)\rangle$, need not satisfy $y(0) = \bar{y}(0)$.

$$H = \{(a, b) \text{ in } R \times R^n : a > 0, b \gg 0\}. \tag{6.3}$$

Clearly, G and H are non-empty, convex sets, and H has a non-empty interior. Also, G and H are disjoint. For if (a, b) belongs to both G and H, then there is some program $\langle x(t), y(t)\rangle$ such that

$$\sum_{t=0}^{\infty} A^t c(t) \leqslant \sum_{t=0}^{\infty} A^t \bar{c}(t) \tag{6.4}$$

and

$$\sum_{t=0}^{\infty} \delta^t w(c(t)) > \sum_{t=0}^{\infty} \delta^t w(\bar{c}(t)). \tag{6.5}$$

Since $\langle \bar{x}(t), \bar{y}(t)\rangle$ is optimal, it is efficient. Hence, by Result 1, $\sum_{t=0}^{\infty} A^t \bar{c}(t) = \tilde{y}$, and (6.4) implies that $\sum_{t=0}^{\infty} A^t c(t) \leqslant \tilde{y}$. So, by Result 1, there is a program $\langle x'(t), y'(t)\rangle$ from $\tilde{y}$, with $c'(t) = c(t)$ for $t \geqslant 0$. In view of (6.5), $\langle \bar{x}(t), \bar{y}(t)\rangle$ is then not optimal, a contradiction. This proves that G and H are disjoint.

Using Theorem 3.5 of Nikaido [10, p. 35], we have (Q, P) in $R_+ \times R_+^n$ such that

$$Qa + Pb \leqslant 0 \qquad \text{for all } (a, b) \text{ in } G. \tag{6.6}$$

We claim now that $Q > 0$. If not, then $Q = 0$, and $P > 0$, and (6.6) implies that

$$Pb \leqslant 0 \qquad \text{for all } (a, b) \text{ in } G. \tag{6.7}$$

Define $\langle x(t), y(t)\rangle$ by $y(t) = \frac{1}{2}\bar{y}(t), x(t) = Ay(t+1)$ for $t \geqslant 0$. Then $\langle x(t), y(t)\rangle$ is a program from $[\tilde{y}/2]$. Now, by Result 1, $\sum_{t=0}^{\infty} A^t \bar{c}(t) = \tilde{y}$, and $\tilde{y} \gg 0$. So, choosing $b = \sum_{t=0}^{\infty} A^t \bar{c}(t) - \sum_{t=0}^{\infty} A^t c(t) \geqslant [\tilde{y}/2] \gg 0$ and $a = \sum_{t=0}^{\infty} \delta^t w(c(t)) - \sum_{t=0}^{\infty} \delta^t w(\bar{c}(t))$, we note that (a, b) is in G, and $Pb > 0$, contradicting (6.7). Hence $Q > 0$. Define $p = P/Q$, and note that $p \geqslant 0$; also, using (6.6)

$$a + pb \leqslant 0 \qquad \text{for all } (a, b) \text{ in } G. \tag{6.8}$$

Thus, given any program $\langle x(t), y(t)\rangle$, we have

$$\sum_{t=0}^{\infty} \delta^t w(c(t)) - p \sum_{t=0}^{\infty} A^t c(t) \leqslant \sum_{t=0}^{\infty} \delta^t w(\bar{c}(t)) - p \sum_{t=0}^{\infty} A^t \bar{c}(t). \tag{6.9}$$

Now, given any c in R_+^n, and $s \geqslant 0$, define $c(t) = \bar{c}(t)$ for $t \neq s$, $c(t) = c$ for $t = s$. Then, there is a program $\langle x'(t), \bar{y}(t)\rangle$ from $\sum_{t=0}^{\infty} A^t c(t)$, with $c'(t) = c(t)$ for $t \geqslant 0$, by Result 1. Using this in (6.9), we obtain

$$\delta^s w(c(s)) - pA^s c \leqslant \delta^s w(\bar{c}(s)) - pA^s \bar{c}(s). \tag{6.10}$$

Define $\bar{p}(t)=pA^t$ for $t\geqslant 0$. Then (6.10) implies

$$\delta^t w(\bar{c}(t))-\bar{p}(t)\,\bar{c}(t)\geqslant\delta^t w(c)-\bar{p}(t)\,c \qquad \text{for all } c \text{ in } R^n_+. \tag{6.11}$$

Also, for $t\geqslant 0$, we have $\bar{p}(t+1)\,\bar{y}(t+1)-\bar{p}(t)\,\bar{x}(t)=\bar{p}(t+1)\,\bar{y}(t+1)-\bar{p}(t)\,A\bar{y}(t+1)$ [since $\bar{x}(t)=A\bar{y}(t+1)$ by efficiency of $\langle\bar{x}(t),\bar{y}(t)\rangle$] $=0$. And, for $t\geqslant 0$, and any (x,y) in Ω, $\bar{p}(t+1)\,y-\bar{p}(t)\,x\leqslant\bar{p}(t+1)\,y-\bar{p}(t)\,Ay=0$. Thus, for $t\geqslant 0$,

$$\bar{p}(t+1)\,\bar{y}(t+1)-\bar{p}(t)\,\bar{x}(t)\geqslant\bar{p}(t+1)\,y-\bar{p}(t)\,x \qquad \text{for all } (x,y) \text{ in } \Omega. \tag{6.12}$$

Clearly (6.11), (6.12) show that $\langle\bar{x}(t),\bar{y}(t),\bar{p}(t)\rangle$ is a competitive program.

Finally, note that for $T\geqslant 1$,

$$\begin{aligned} p\left[\sum_{t=0}^{T} A^t\bar{c}(t)\right] &= p\left[\sum_{t=0}^{T} A^t[\bar{y}(t)-\bar{x}(t)]\right] \\ &= p\left[\bar{y}(0)+\sum_{t=0}^{T-1} A^{t+1}\bar{y}(t+1)-\sum_{t=0}^{T-1} A^t\bar{x}(t)-A^T\bar{x}(T)\right] \\ &= p[\bar{y}(0)-A^T\bar{x}(T)]. \end{aligned}$$

Hence,

$$\begin{aligned} p\sum_{t=0}^{\infty} A^t\bar{c}(t) &= p\bar{y}(0)-\lim_{t\to\infty} pA^t\bar{x}(t) \\ &= p\bar{y}(0)-\lim_{t\to\infty}\bar{p}(t)\,\bar{x}(t). \end{aligned}$$

Since $\sum_{t=0}^{\infty} A^t\bar{c}(t)=\bar{y}(0)$ by efficiency, so $p\sum_{t=0}^{\infty} A^t\bar{c}(t)=p\bar{y}(0)$. Hence $\lim_{t\to\infty}\bar{p}(t)\,\bar{x}(t)=0$, which proves the proposition. ∎

Remark. It is worth noting that the price sequence $\langle\bar{p}(t)\rangle$ will, in fact, satisfy $\bar{p}(t)\gg 0$ for $t\geqslant 0$ (see Lemma 2 below). So the price, p, obtained from the separation theorem must satisfy $p>0$.

The proof of Proposition 5 (the existence of a steady-state optimal program) requires a sequence of preliminary results which we now discuss. First, using the assumption on the utility function [(A.3)], it is easy to prove the following result.

LEMMA 1. (a) *Suppose p^0 is in R^n_+. Then there exists $\theta^0>0$ such that*

$$w(\theta e)-p^0(\theta e)>0 \qquad \textit{for} \quad 0<\theta\leqslant\theta^0. \tag{6.13}$$

(b) *Suppose* p^0 *and* c^0 *are in* R^n_+, *respectively. If* $w(c^0)-p^0c^0 \geqslant w(c)-p^0c$ *for all* c *in* R^n_+, *then* $w(c^0)-p^0c^0>0$, $w(c^0)>0$, $c^0>0$.

Then, using Lemma 1, one can establish that competitive programs are "interior" in input–output levels.

LEMMA 2. *Suppose* $\langle x(t), y(t), p(t)\rangle$ *is a competitive program from* $\tilde{y}$ *in* R^n_+. *Then* (i) $w(c(t))>0$, $c(t)>0$, $x(t))\gg 0$, $y(t)\gg 0$, *and* $p(t)\gg 0$; (ii) $p(t+1)=p(t)\,A$ *for* $t\geqslant 0$ *and* $Ay(t+1)=x(t)$ *for* $t\geqslant 0$.

Proof. The only non-trivial part of (i) is to show that $p(t)\gg 0$ for $t\geqslant 0$. Using (2.1) we have for $t\geqslant 0$, $i=1,...,n$, $\delta^t w(c(t))-p(t)\,c(t)\geqslant \delta^t w(c(t)+e^i)-p(t)(c(t)+e^i)$. Hence, for $i=1,...,n$, $t\geqslant 0$, $p_i(t)=p(t)\,e^i\geqslant \delta^t[w(c(t)+e^i)-w(c(t))]$. Since $w(c(t))>0$, we have $f(c(t))>0$ and by (A.3) (c), $f(c(t)+e^i)>f(c(t))$. Therefore, $w(c(t)+e^i)>w(c(t))$, and $p_i(t)>0$.

Part (ii) follows from Result 3(iii) and 3(iv), and (2.2), since $y(t)\gg 0$ and $p(t)\gg 0$ for $t\geqslant 0$ by part (i) above. ∎

Lemma 2 in turn yields the result that positive scalar multiples of competitive programs are also competitive programs.

LEMMA 3. (a) *If* c^0 *in* R^n_+ *and* p^0 *in* R^n_+ *are such that*, $w(c^0)-p^0c^0\geqslant w(c))-p^0c$ *for all* c *in* R^n_+, *then for any* $\beta>0$, $w(\hat{\beta}c^0)-\beta p^0(\hat{\beta}c^0)\geqslant w(c)-\beta p^0c$ *for all* c *in* R^n_+, *where* $\hat{\beta}\equiv[1/\beta^{1/\alpha}]$.

(b) *If* $\langle x(t), y(t), p(t)\rangle$ *is a competitive progam from* $\tilde{y}$ *with consumption sequence* $\langle c(t)\rangle$, *then for any* $\beta>0$, $\langle \beta x(t), \beta y(t), \beta^{-\alpha}p(t)\rangle$ *is a competitive program from* $\beta\tilde{y}$, *with consumption sequence* $\langle \beta c(t)\rangle$.

Proof. Suppose c^0 and p^0 are in R^n_+ and $w(c^0)-p^0c^0\geqslant w(c)-p^0c$ for all c in R^n_+. Consider any $\beta>0$ and any c in R^n_+. Then

$$\begin{aligned} w(\hat{\beta}c^0)-\beta p^0(\hat{\beta}c^0) &= \beta\hat{\beta}[w(c^0)-p^0c^0] \\ &\geqslant \beta\hat{\beta}[w(c/\hat{\beta})-p^0(c/\hat{\beta})]=w(c)-\beta p^0c. \end{aligned}$$

This proves (a). To prove (b), note that by Lemma 2, $x(t)=Ay(t+1)$ for $t\geqslant 0$. Hence, $\beta x(t)=A\beta y(t+1)$. Since $0\leqslant c(t)=y(t)-x(t)$ for $t\geqslant 0$, therefore, $0\leqslant \beta c(t)=\beta y(t)-\beta x(t)$ for $t\geqslant 0$. Hence, $\langle \beta x(t), \beta y(t)\rangle$ is a program with consumption sequence $\langle \beta c(t)\rangle$. To see that $\langle \beta x(t), \beta y(t), \beta^{-\alpha}p(t)\rangle$ is competitive, we note that clearly $\beta^{-\alpha}p(t)\geqslant 0$, and $w(\beta c(t))-\beta^{-\alpha}p(t)(\beta c(t))\geqslant w(c)-\beta^{-\alpha}p(t)\,c$ for any c in R^n_+, by part (a) above. We need to show that $\beta^{-\alpha}p(t+1)\ (\beta y(t+1))-\beta^{-\alpha}p(t)(\beta x(t))\geqslant$

$\beta^{-\alpha}p(t+1)\,y-\beta^{-\alpha}p(t)\,x$ for any (x, y) in Ω, $t \geqslant 0$. This follows from (2.2); that is,

$$p(t+1)\,y(t+1)-p(t)\,x(t) \geqslant p(t+1)\,y-p(t)\,x \qquad \text{for all } (x, y) \text{ in } \Omega,\ t \geqslant 0,$$

which in turn implies that for any (x, y) in Ω, and $t \geqslant 0$,

$$\begin{aligned} &\beta^{-\alpha}p(t+1)\,\beta y(t+1)-\beta^{-\alpha}p(t)\,\beta x(t) \\ &\quad \geqslant \beta^{-\alpha}p(t+1)\,\beta(y/\beta)-\beta^{-\alpha}p(t)\,\beta(x/\beta) \\ &\quad \left[\text{since } (x, y) \text{ is in } \Omega \text{ implies } \left(\frac{1}{\beta}x, \frac{1}{\beta}y\right) \text{ is in } \Omega\right] \\ &\quad = \beta^{-\alpha}p(t+1)\,y-\beta^{-\alpha}p(t)\,x \end{aligned}$$

which completes the proof. ▮

Using Lemma 3 now yields the existence result on a steady-state optimal program (Proposition 5) which we now prove.

Proof of Proposition 5

We first define a set

$$S=\{c \text{ in } R^n_+ : \|c\| \leqslant [w(e)/m(\hat{p})]^{1/\alpha}\}.$$

We note that for c in R^n_+, c not in S, we have

$$w(c)-\hat{p}c<0. \tag{6.14}$$

To see this note that for c not in S,

$$\begin{aligned} w(c)-\hat{p}c &= \|c\|[\{w(c)/\|c\|\}-\{\hat{p}c/\|c\|\}] \\ &\leqslant \|c\|[\{w(c/\|c\|)/\|c\|^{\alpha}\}-m(\hat{p})] \\ &\leqslant \|c\|[w(e)/\|c\|^{\alpha}-m(\hat{p})] \\ &= \|c\|^{1-\alpha}\,m(\hat{p})[w(e)/m(\hat{p})-\|c\|^{\alpha}] \\ &<0, \qquad \text{since } \|c\|>[w(e)/m(\hat{p})]^{1/\alpha}. \end{aligned}$$

We also note that, by Lemma 1, there exists θ' such that

$$w(\theta' e)-\hat{p}(\theta' e)>0 \qquad \text{and} \qquad 0<\theta'<[w(e)/\|\hat{p}\|]^{1/\alpha}/n. \tag{6.15}$$

Now, S is a non-empty, compact set in R^n_+, and the function, $F(c) \equiv w(c)-\hat{p}c$ for c in R^n_+, is continuous on S. So, there is c^* in S, such that

$$w(c^*)-\hat{p}c^* \geqslant w(c)-\hat{p}c \qquad \text{for all } c \text{ in } S. \tag{6.16}$$

Let $c' = \theta' e$. Then $\| c' \| = n\theta' < [w(e)/\| \hat{p} \|]^{1/\alpha} \leqslant [w(e)/m(\hat{p})]^{1/\alpha}$, so c' is in S. Using this in (6.16), we have $w(c^*) - \hat{p}c^* \geqslant w(c') - \hat{p}c' > 0$ [by (6.15)]. This also implies $c^* \dot{>} 0$, by Remark 1 in Section 2b. Since for c not in S, $w(c) - \hat{p}c < 0$, so by (6.16)

$$w(c^*) - \hat{p}c^* \geqslant w(c) - \hat{p}c \qquad \text{for all } c \text{ in } R^n_+. \tag{6.17}$$

Define $g = (\delta\hat{\lambda})^{1/\alpha}$, and note that by (A.4), we have

$$g < \left[\left(\frac{1}{\hat{\lambda}^{1-\alpha}}\right)\hat{\lambda}\right]^{1/\alpha} = [\hat{\lambda}^\alpha]^{1/\alpha} = \hat{\lambda} \tag{6.18}$$

so that, by the Frobenius theorem, we have $(I - gA)$ is non-singular (invertible), and $(I - gA)^{-1} \gg 0$ (Nikaido [10, p. 102 and p. 107]). Define

$$y^* = c^*(I - gA)^{-1}. \tag{6.19}$$

Now, define the sequence $\langle x^*(t), y^*(t)\rangle$ from y^* by $y^*(0) = y^*$, $y^*(t+1) = gy^*(t)$ for $t \geqslant 0$; $x^*(t) = Ay^*(t+1)$ for $t \geqslant 0$. Note, then, that $(x^*(t), y^*(t+1))$ is in Ω for $t \geqslant 0$. Also $y^*(t) + x^*(t) = y^*(t) - Ay^*(t+1) = g^t y^* - Ag^{t+1}y^* = g^t[I - gA]y^* = c^* g^t$ [by (6.19)] $\geqslant 0$. Hence $\langle x^*(t), y^*(t)\rangle$ is a program from y^*, and $c^*(t) = c^* g^t$ for $t \geqslant 0$ is the corresponding consumption sequence. Note that $c^* > 0$, $(I - gA)^{-1} \gg 0$ implies that $y^* \gg 0$ and hence $x^* \gg 0$.

Clearly, $\langle x^*(t), y^*(t)\rangle$ is a steady-state program. Our next task is to show that it is also an optimal program. To this end, define

$$p^*(t) = [\hat{p}/\hat{\lambda}^t] \qquad \text{for} \quad t \geqslant 0. \tag{6.20}$$

We will now show that for $t \geqslant 0$,

(i) $\delta^t w(c^*(t)) - p^*(t)\, c^*(t) \geqslant \delta^t w(c) - p^*(t)\, c$ for all c in R^n_+

(ii) $0 = p^*(t+1)\, y^*(t+1) - p^*(t)\, x^*(t) \geqslant p^*(t+1)\, y - p^*(t)\, x$ for all (x, y) in Ω.

To establish (i), note that by (6.17) and Lemma 3(a) we have

$$w(g^t c^*) - (1/(g^t)^\alpha)\, \hat{p}(g^t c^*) \geqslant w(c) - (1/(g^t)^\alpha)\, \hat{p}c \qquad \text{for all } c \text{ in } R^n_+,\ t \geqslant 0.$$

Therefore, using $g^\alpha = \delta\hat{\lambda}$ and $c^*(t) = g^t c^*$ for $t \geqslant 0$,

$$w(c^*(t)) - (1/\delta^t\hat{\lambda}^t)\, \hat{p}c^*(t) \geqslant w(c) - (1/\delta^t\hat{\lambda}^t)\, \hat{p}c \qquad \text{for all } c \text{ in } R^n_+,\ t \geqslant 0,$$

which is (i).

To establish (ii) note that for $t \geqslant 0$, $p^*(t+1) - p^*(t)\, A = (1/\hat{\lambda}^{t+1})(\hat{p} - \lambda \hat{p} A) = 0$. Therefore, for any (x, y) in Ω and $t \geqslant 0$ we have

$$\begin{aligned} p^*(t+1)\, y - p^*(t)\, x &\leqslant p^*(t+1)\, y - p^*(t)\, Ay = 0 \\ &= p^*(t+1)\, y^*(t+1) - p^*(t)\, Ay^*(t+1) \\ &= p^*(t+1)\, y^*(t+1) - p^*(t)\, x^*(t). \end{aligned}$$

Finally, note that $p^*(t)\, y^*(t) = (\hat{p}/\hat{\lambda}^t)\ y^* g^t = \hat{p} y^*\, (g/\hat{\lambda})^t$. Since $0 < (g/\hat{\lambda}) < 1$ [by (6.18)], so $p^*(t)\, y^*(t) \to 0$ as $t \to \infty$; that is, the transversality condition is satisfied. Hence $\langle x^*(t), y^*(t) \rangle$ is optimal from y^* by Proposition 1. Since it is a steady-state program, so it is a steady-state optimal program. ▮

The proof of Proposition 6 (the turnpike property of optimal programs) requires several preliminary results, which we now discuss. First, using Lemma 3 above, it is easy to prove the following result.

LEMMA 4. *Suppose* $\langle x(t), y(t), p(t) \rangle$ *is a competitive program from* $\tilde{y}$ *in* R^n_+. *Then*

(i) $w(c(t)/g^t) - \hat{\lambda}^t p(t)(c(t)/g^t) \geqslant w(c) - \hat{\lambda}^t p(t)\, c$ *for all* c *in* R^n_+, $t \geqslant 0$;

and

(ii) $w(c(t)/g^t) - \hat{\lambda}^t p(t)(c(t)/g^t) > 0$ *for* $t \geqslant 0$.

The additional assumption on the welfare function [(A.5)] effectively makes it strictly concave everywhere in the part of the domain of w which is of interest, namely where $w(c) > 0$. This is what the following Lemma establishes.

LEMMA 5. *Under* (A.3) *and* (A.5), *if* c^0, c^1 *are in* R^n_+, $c^0 \neq c^1$, $w(c^1) > 0 < w(c^0)$, *and* $0 < \theta < 1$, *then*

$$w(\theta c^0 + (1-\theta)\, c^1) > \theta w(c^0) + (1-\theta)\, w(c^1).$$

Proof. If c^0 is not proportional to c^1, then,

$$\begin{aligned} & w(\theta c^0 + (1-\theta)\, c^1) \\ &\quad = [f(\theta c^0 + (1-\theta)\, c^1)]^{1-\alpha} \\ &\quad \geqslant [\theta f(c^0) + (1-\theta) f(c^1)]^{1-\alpha} \qquad \text{[by (A.5)]} \\ &\quad \geqslant \theta (f(c^0))^{1-\alpha} + (1-\theta)(f(c^1))^{1-\alpha} \\ &\qquad \text{[since } x^{1-\alpha} \text{ is concave for } x \geqslant 0] \\ &\quad = \theta w(c^0) + (1-\theta)\, w(c^1). \end{aligned}$$

If c^0 is proportional to c^1, then there exists $\mu \neq 0$ such that $c^0 = \mu c^1$. Since c^0, c^1 are in R^n_+, therefore, $\mu > 0$. Since $c^0 \neq c^1$, therefore, $\mu \neq 1$. Therefore,

$$\begin{aligned}
&w(\theta c^0 + (1-\theta)\, c^1) \\
&\quad = w((\theta\mu + 1 - \theta)\, c^1) \\
&\quad = (\theta\mu + 1 - \theta)^{1-\alpha}\, w(c_1) \\
&\quad > [\theta\mu^{1-\alpha} + (1-\theta)\, 1^{1-\alpha}]\, w(c_1) \\
&\qquad [\text{since, } x^{1-\alpha} \text{ is a strictly concave function of } x > 0, \\
&\qquad \mu \neq 1, \mu > 0 \text{ and } 0 < \theta < 1] \\
&\quad = \theta\mu^{1-\alpha} w(c^1) + (1-\theta)\, w(c^1) \\
&\quad = \theta w(\mu c^1) + (1-\theta)\, w(c^1) \\
&\quad = \theta w(c^0) + (1-\theta)\, w(c^1).
\end{aligned}$$

This establishes the lemma. ∎

Since the welfare function is strictly concave in the relevant domain, therefore, convergence of the *supporting prices* of a sequence of consumption points to that of a given point implies the convergence of the sequence of consumption points to the given point. This is the content of the following lemma.

LEMMA 6. *Under* (A.3) *and* (A.5), *if* $\langle p^s \rangle$, $\langle c^s \rangle$ *are sequences in* R^n_+, $\bar{p}$ *and* $\bar{c}$ *are in* R^n_+ *and*

(i) $\lim_{s\to\infty} p^s = \bar{p}$,

(ii) $w(c^s) - p^s c^s \geqslant w(c) - p^s c$ *for all* c *in* R^n_+, *for each* $s \geqslant 0$,

(iii) $w(\bar{c}) - \bar{p}\bar{c} \geqslant w(c) - \bar{p}c$ *for all* c *in* R^n_+,

then $\lim_{s\to\infty} c^s = \bar{c}$.

Proof. First note that, by Lemma 1(b), (ii) and (iii), respectively, imply that

$$w(c^s) > 0 \qquad \text{for} \quad s \geqslant 0; \qquad \text{and} \qquad w(\bar{c}) > 0. \tag{6.21}$$

Now, suppose the lemma is false. Then, without any loss of generality, we may suppose that there exists $\varepsilon_0 > 0$ such that

$$\| c^s - \bar{c} \| > \varepsilon_0 \qquad \text{for} \quad s \geqslant 0 \tag{6.22}$$

and [by virtue of the continuity of w and (6.21) above] that

$$w(c) > 0, \qquad \text{for } c \text{ in } R^n_+ \text{ satisfying } \| c - \bar{c} \| \leqslant \varepsilon_0. \tag{6.23}$$

Let $B = \{c \text{ in } R^n_+ : \| c - \bar{c} \| = \varepsilon_0\}$. Clearly B is compact. Consider any c in B. Then by Lemma 5, we have

$$w((\bar{c}/2) + (c/2)) > \tfrac{1}{2}(w(\bar{c}) + w(c))$$

[since $c \neq \bar{c}$ and $w(c) > 0$, $0 < w(\bar{c})$ and, therefore, Lemma 5 applies]. Therefore, $2[w((\bar{c}/2) + (c/2)) - w(\bar{c})] - [w(c) - w(\bar{c})] > 0$ for c in B. Since B is compact and w is continuous, therefore, there exists $\varepsilon_1 > 0$ such that

$$2[w((\bar{c}/2) + (c/2)) - w(\bar{c})] - [w(c) - w(\bar{c})] \geqslant \varepsilon_1 \qquad \text{for } c \text{ in } B. \tag{6.24}$$

Now, consider any $s \geqslant 0$. Since $\| c^s - \bar{c} \| > \varepsilon_0$, therefore, there exist $\lambda^1 > 1$ and c^1 in B such that $(1/\lambda^1)\, c^s + (1 - (1/\lambda^1))\, \bar{c} = c^1$. Since $w(c^s) > 0$, $w(\bar{c}) > 0$, $c^s \neq \bar{c}$, and $0 < (1/\lambda^1) < 1$, therefore, by Lemma 5, $w(c^1) > (1/\lambda^1)\, w(c^s) + (1 - (1/\lambda^1))\, w(\bar{c})$. Since $\lambda^1 > 0$, therefore, $\lambda^1[w(c^1) - w(\bar{c})] > w(c^s) - w(\bar{c})$. Therefore,

$$(\lambda^1 - 1)[w(c^1) - w(\bar{c})] > w(c^s) - w(c^1). \tag{6.25}$$

Now, from (ii), substituting c^1 for c, we obtain

$$\begin{aligned} w(c^1) - p^s c^1 &\leqslant w(c^s) - p^s c^s = w(c^s) - p^s[\lambda^1 c^1 + (1 - \lambda^1)\, \bar{c}] \\ &= w(c^s) - p^s \bar{c} - \lambda^1 p^s (c^1 - \bar{c}). \end{aligned}$$

Therefore,

$$(\lambda^1 - 1)\, p^s (c^1 - \bar{c}) \leqslant w(c^s) - w(c^1) < (\lambda^1 - 1)[w(c^1) - w(\bar{c})] \quad [\text{from (6.25)}].$$

Since $\lambda^1 - 1 > 0$, therefore,

$$p^s (c^1 - \bar{c}) < w(c^1) - w(\bar{c}). \tag{6.26}$$

Also, from (iii), substituting $[(\bar{c}/2) + (c^1/2)]$ for c, we obtain

$$w(\bar{c}) - \bar{p}\bar{c} \geqslant w((\bar{c}/2) + (c^1/2)) - \bar{p}((\bar{c}/2) + (c^1/2)).$$

Hence,

$$\bar{p}(c^1 - \bar{c}) \geqslant 2[w((\bar{c}/2) + (c^1/2)) - w(\bar{c})]. \tag{6.27}$$

From (6.26) and (6.27) we obtain

$$\begin{aligned} (\bar{p} - p^s)(c^1 - \bar{c}) &> 2[w((\bar{c}/2) + (c^1/2)) - w(\bar{c})] - [w(c^1) - w(\bar{c})] \\ &\geqslant \varepsilon_1 \qquad [\text{from (6.24), since } c^1 \text{ is in } B]. \end{aligned}$$

Therefore, $0<\varepsilon_1 \leqslant \| \bar{p}-p^s \| \, \| c^1-\bar{c} \| = \| \bar{p}-p^s \| \, \varepsilon_0$. Hence, for $s \geqslant 0$, $\| \bar{p}-p^s \| \geqslant \varepsilon_1/\varepsilon_0 > 0$. This contradicts (i) and, therefore, completes the proof of the lemma. ∎

Lemma 6 now helps us to establish the turnpike property of optimal programs (Proposition 6) under the maintained assumptions (A.1)–(A.5).

Proof of Proposition 6

Since $\langle \bar{x}(t), \bar{y}(t) \rangle$ is an optimal program from $\tilde{y} \gg 0$, there is (by Proposition 3) a sequence $\langle \bar{p}(t) \rangle$ such that $\langle \bar{x}(t), \bar{y}(t), \bar{p}(t) \rangle$ is a competitive program.

By Lemma 2, $[\bar{x}(t), \bar{y}(t), w(\bar{c}(t)), \bar{p}(t)] \gg 0$ for $t \geqslant 0$, and also for $t \geqslant 0$

$$0 = \bar{p}(t+1) - \bar{p}(t)\, A \qquad \text{for} \quad t \geqslant 0. \tag{6.28}$$

From (6.28) it follows that

$$\hat{\lambda}^t \bar{p}(t) = \bar{p}(0)\, \hat{\lambda}^t A^t \qquad \text{for} \quad t \geqslant 0. \tag{6.29}$$

Now, $\hat{\lambda}^t A^t$ converges to a matrix $\bar{A}$, such that $\bar{a}_{ij} = \hat{p}_j \, \hat{y}_i$ (Karlin [7, p. 249]) for $i=1, ..., n; j=1, ..., n$. Thus, we have

$$\hat{\lambda}^t \bar{p}(t) \to [\bar{p}(0)\, \hat{y}]\, \hat{p} \qquad \text{as} \quad t \to \infty. \tag{6.30}$$

Define $\mu = 1/[\bar{p}(0)\, \hat{y}]^{1/\alpha}$, and a sequence $\langle x'(t), y'(t) \rangle$ by $y'(t) = \mu y^*(t)$, $x'(t) = \mu x^*(t)$ for $t \geqslant 0$, where $\langle x^*(t), y^*(t) \rangle$ is the steady-state optimal program of Proposition 5. Also, define $\langle p'(t) \rangle$ by $p'(t) = \mu^\alpha p^*(t)$ for $t \geqslant 0$ and $\langle c'(t) \rangle$ by $c'(t) = \mu c^*(t)$ for $t \geqslant 0$, where $c^*(t)$ is the consumption sequence associated with $\langle x^*(t), y^*(t) \rangle$ and $\langle p^*(t) \rangle$ is the corresponding sequence of present value prices. Now $\langle x^*(t), y^*(t), p^*(t) \rangle$ is competitive, and $\mu > 0$ since $\bar{p}(0) \gg 0$ by Lemma 2. Therefore, by Lemma 3(b), $\langle x'(t), y'(t), p'(t) \rangle$ is competitive. Moreover, $\lim_{t \to \infty} p'(t)\, x'(t) = \lim_{t \to \infty} \mu^{1-\alpha} p^*(t)\, x^*(t) = 0$. Hence $\langle x'(t), y'(t) \rangle$ is optimal. Clearly it is a steady-state program with consumption sequence $\langle c'(t) \rangle$. Therefore $\langle x'(t), y'(t) \rangle$ is a steady-state optimal program. This establishes (i).

Now, $p'(0) = \mu^{-\alpha} p^*(0) = [\bar{p}(0)\, \hat{y}]\, \hat{p}$. Therefore, from (6.30), we have

$$\hat{\lambda}^t \bar{p}(t) \to p'(0) \qquad \text{as} \quad t \to \infty. \tag{6.31}$$

Since $\langle x'(t), y'(t), p'(t) \rangle$ is competitive, we have

$$w(c'(0)) - p'(0)\, c'(0) \geqslant w(c) - p'(0)\, c \qquad \text{for all } c \text{ in } R^n_+. \tag{6.32}$$

Since $\langle \bar{x}(t), \bar{y}(t), \bar{p}(t) \rangle$ is competitive, Lemma 4 yields

$$w(\bar{c}(t)/g^t) - \hat{\lambda}^t \bar{p}(t)(\bar{c}(t)/g^t) \geqslant w(c) - \hat{\lambda}^t \bar{p}(t)\, c \qquad \text{for all } c \text{ in } R^n_+,\ t \geqslant 0. \tag{6.33}$$

By virtue of (6.31), (6.32), and (6.33) we may appeal to Lemma 6 and conclude that

$$\lim_{t \to \infty} (\bar{c}(t)/g^t) = c'(0). \tag{6.34}$$

Now, in view of the "principle of optimality," for $s \geqslant 0$, we have $\langle \bar{x}(t+s), \bar{y}(t+s) \rangle$ is optimal from $\bar{y}(s)$. So, for $s \geqslant 0$, we have $\langle \bar{x}(t+s), \bar{y}(t+s) \rangle$ is efficient from $\bar{y}(s)$. Hence, using Result 1, we have, for $s \geqslant 0$,

$$\bar{y}(s) = \sum_{t=0}^{\infty} A^t \bar{c}(s+t) = \sum_{t=0}^{\infty} g^t A^t [\bar{c}(s+t)/g^t].$$

So, we obtain for $s \geqslant 0$,

$$[\bar{y}(s)/g^s] = \sum_{t=0}^{\infty} g^t A^t [\bar{c}(s+t)/g^{t+s}].$$

For the steady-state optimal program $\langle x'(t), y'(t) \rangle$, we have

$$y'(0) = \mu y^* = \mu(I-gA)^{-1} c^* = (I-gA)^{-1} c'(0) = \sum_{t=0}^{\infty} g^t A^t c'(0).$$

So, we obtain, for $s \geqslant 0$,

$$[\{\bar{y}(s)/g^s\} - y'(0)] = \sum_{t=0}^{\infty} g^t A^t [\{\bar{c}(s+t)/g^{t+s}\} - c'(0)]. \tag{6.35}$$

Now, given any $\varepsilon > 0$, there is, by (6.34), a positive integer N^*, such that $N \geqslant N^*$ implies $\| \{\bar{c}(N)/g^N\} - c'(0) \| \leqslant \varepsilon$. Using this in (6.35), we have for $s \geqslant N^*$

$$[\{\bar{y}(s)/g^s\} - y'(0)] \leqslant \varepsilon \sum_{t=0}^{\infty} g^t A^t e = \varepsilon (I-gA)^{-1} e \qquad [\text{since } gA \geqslant 0]. \tag{6.36}$$

Similarly, by (6.35), for $s \geqslant N^*$,

$$[\{\bar{y}(s)/g^s\} - y'(0)] \geqslant -\varepsilon \sum_{t=0}^{\infty} g^t A^t e = -\varepsilon (I-gA)^{-1} e. \tag{6.37}$$

Thus, for $s \geqslant N^*$,

$$\| \{\bar{y}(s)/g^s\} - y'(0) \| \leqslant \varepsilon \, \| (I-gA)^{-1} e \| \tag{6.38}$$

which proves that $\{\bar{y}(s)/g^s\} \to y'(0)$ as $s \to \infty$. This establishes (ii). ∎

Proof of Theorem 1

We will show that (5.1) implies

$$\liminf_{t\to\infty} \bar{p}(t)\,\bar{y}(t)=0 \tag{6.39}$$

so that the optimality of $\langle \bar{x}(t), \bar{y}(t)\rangle$ then follows by Proposition 1.

To this end, our first objective is to show that

$$\bar{p}(0)\,\hat{y} > \hat{p}\hat{y}. \tag{6.40}$$

We start by noting that if y is in Y, then $w(c^*) - w(y/\theta) \geqslant \hat{p}c^* - \hat{p}(y/\theta)$ [using (6.17)] $= 0$ [using the definition of Y]. So $w(c^*) \geqslant w(y/\theta) = w(y)/\theta^{1-\alpha}$ and so $\theta^{1-\alpha}w(c^*) \geqslant w(y)$. Define $a = \theta^{1-\alpha}$, and $[(1/a^\alpha) - 1] = b$. Note that since $0 < \theta < 1$, so $0 < a < 1$, and $b > 0$. Since $\bar{y}(0)$ is in Y, therefore, $aw(c^*) \geqslant w(\bar{y}(0))$.

Now, since $\langle \bar{x}(t), \bar{y}(t)\rangle$ is competitive $w(\bar{c}(0) + a\,\|c^*\|\,e) - w(\bar{c}(0)) \leqslant \bar{p}(0)\,a\,\|c^*\|\,e = a\,\|c^*\|\,\|\bar{p}(0)\|$. Also, using the fact that $\bar{c}(0) \leqslant \bar{y}(0)$, and so $w(\bar{c}(0)) \leqslant w(\bar{y}(0))$, we have $w(\bar{c}(0) + a\,\|c^*\|\,e) - w(\bar{c}(0)) \geqslant w(a\,\|c^*\|\,e) - w(\bar{y}(0)) \geqslant w(ac^*) - aw(c^*) = a^{1-\alpha}w(c^*) - aw(c^*) = a[(1/a^\alpha) - 1]\,w(c^*) = ab\,w(c^*)$. Using the above two results, we conclude that

$$\|\bar{p}(0)\| \geqslant bw(c^*)/\|c^*\|. \tag{6.41}$$

This yields $\bar{p}(0)\,\hat{y} \geqslant m(\hat{y})\,\bar{p}(0)\,e = m(\hat{y})\|\bar{p}(0)\| \geqslant m(\hat{y})\,bw(c^*)/\|c^*\| > m(\hat{y})\,b\hat{p}c^*/\|c^*\|$ [using (6.15) and (6.17)] $\geqslant m(\hat{y})\,b\hat{p}\hat{y}/E\,\|c^*\|$ (using the definition of E). Summarizing these inequalities, we have

$$\bar{p}(0)\,\hat{y} > (\hat{p}\hat{y})\,[bm(\hat{y})/E\,\|c^*\|] = \hat{p}\hat{y}b\eta. \tag{6.42}$$

Using the definition of θ, $\theta^{\alpha(1-\alpha)} = \eta/[\eta+1]$, so that we have $[1/\theta^{\alpha(1-\alpha)}] = 1 + (1/\eta)$. This means that $\{[1/\theta^{\alpha(1-\alpha)}] - 1\}\,\eta = 1$; or, using the notation $a = \theta^{1-\alpha}$, and $[(1/a^\alpha) - 1] = b$, we have $b\eta = 1$. Using this last piece of information in (6.42) yields (6.40).

Now, suppose, contrary to (6.39), that $\liminf_{t\to\infty} \bar{p}(t)\,\bar{y}(t) > 0$. Then, there is $\mu > 0$, and an integer $T_1 \geqslant 0$, such that $t \geqslant T_1$ implies

$$\bar{p}(t)\,\bar{y}(t) \geqslant \mu. \tag{6.43}$$

Since $\langle \bar{x}(t), \bar{y}(t), \bar{p}(t)\rangle$ is competitive, therefore, by Lemma 2, we have

$$\bar{p}(t+1) = \bar{p}(t)\,A \qquad \text{for} \quad t \geqslant 0. \tag{6.44}$$

Using (6.44), we note that for $t \geqslant 0$,

$$\hat{\lambda}^{t+1}\bar{p}(t+1)\,\hat{y} = \hat{\lambda}^{t+1}\bar{p}(t)\,A\hat{y} = \hat{\lambda}^{t}\bar{p}(t)\,\hat{y}$$

and so we have

$$\hat{\lambda}^t \bar{p}(t)\,\hat{y} = \bar{p}(0)\,\hat{y}. \tag{6.45}$$

Similarly, using (6.20), we have

$$\hat{\lambda}^t p^*(t)\,\hat{y} = \hat{p}\hat{y}. \tag{6.46}$$

Denoting $[\bar{p}(0)\,\hat{y} - \hat{p}\hat{y}]$ by ε, and noting by (6.40) that $\varepsilon > 0$, we have, by (6.45) and (6.46),

$$\hat{\lambda}^t [\bar{p}(t) - p^*(t)]\,\hat{y} = \varepsilon. \tag{6.47}$$

Now, in view of (6.44), we know that

$$\hat{\lambda}^t \bar{p}(t) = \bar{p}(0)\,\hat{\lambda}^t A^t \qquad \text{for} \quad t \geqslant 0. \tag{6.48}$$

Since $\hat{\lambda}^t A^t$ converges to a matrix $\bar{A}$, such that $\bar{a}_{ij} = \hat{p}_j \hat{y}_i$ (Karlin [7, p. 249]), we have

$$\hat{\lambda}^t \bar{p}(t) \to [\bar{p}(0)\,\hat{y}]\,\hat{p} \qquad \text{as} \quad t \to \infty. \tag{6.49}$$

On the other hand, $\hat{\lambda}^t p^*(t) = \hat{p}$; so, using $\hat{p}\hat{y} = 1$ we can write

$$\hat{\lambda}^t p^*(t) = [\hat{p}\hat{y}]\,\hat{p} \qquad \text{for} \quad t \geqslant 0. \tag{6.50}$$

Combining (6.49) and (6.50), we can find $T_2 \geqslant T_1$, such that

$$\hat{\lambda}^t [\bar{p}(t) - p^*(t)] \geqslant (\varepsilon/2)\,\hat{p} \qquad \text{for} \quad t \geqslant T_2. \tag{6.51}$$

Now $y^*(t) = y^* g^t$ for $t \geqslant 0$, where $0 < g < \hat{\lambda}$. So, in view of (6.49), (6.50), we certainly have $[\bar{p}(t) - p^*(t)]\,y^*(t) \to 0$ as $t \to \infty$. Defining $v = [\mu\varepsilon/8\bar{p}(0)\,\hat{y}]$ we can find $T_3 \geqslant T_2$, such that

$$[\bar{p}(t) - p^*(t)]\,y^*(t) \leqslant v \qquad \text{for} \quad t \geqslant T_3. \tag{6.52}$$

Using this in (5.2) we have for $t \geqslant T_3$

$$[\bar{p}(t) - p^*(t)]\,\bar{y}(t) \leqslant v. \tag{6.53}$$

Combining this with (6.51) yields

$$(\varepsilon/2)\,\hat{p}\bar{y}(t)/\hat{\lambda}^t \leqslant v \qquad \text{for} \quad t \geqslant T_3. \tag{6.54}$$

Relying on (6.49) again, we can find $T_4 \geqslant T_3$, such that $t \geqslant T_4$ implies

$$\hat{\lambda}^t \bar{p}(t) \leqslant 2[\bar{p}(0)\,\hat{y}]\,\hat{p}. \tag{6.55}$$

Using this in (6.54), we have, for $t \geqslant T_4$,

$$\bar{p}(t)\,\bar{y}(t) < 2[\bar{p}(0)\,\hat{y}]\,\hat{p}\bar{y}(t)/\hat{\lambda}^t < \{4\bar{p}(0)\,\hat{y}/\varepsilon\}\,v = (\mu/2), \tag{6.56}$$

which contradicts (6.43) and establishes (6.39), and hence the theorem. ∎

References

1. H. Atsumi, The efficient capital program for a maintainable utility level, *Rev. Econ. Stud.* **36** (1969), 263–288.
2. W. A. Brock and D. Gale, Optimal growth under factor augmenting progress, *J. Econ. Theory* **1** (1969), 229–243.
3. W. A. Brock and M. Majumdar, "On Characterizing Optimality of Competitive Programs in Terms of Decentralizable Conditions," Cornell University Working Paper No. 333 (1985).
4. S. Dasgupta and T. Mitra, "Intertemporal Optimality in a Closed Linear Model of Production," Cornell University Working Paper No. 334 (1985).
5. D. Gale, "The Theory of Linear Economic Models," McGraw–Hill, New York, 1958.
6. F. R. Gantmacher, "Matrix Theory," Vol. 2, Chelsea, New York, 1959.
7. S. Karlin, "Mathematical Methods in Games, Programming and Mathematical Economics," Vol. 1, Addison–Wesley, Reading, MA, 1959.
8. D. McFadden, The evaluation of development programs, *Rev. Econ. Stud.* **34** (1967), 25–50.
9. M. Majumdar, Efficient programs in infinite dimensional spaces: A complete characterization, *J. Econ. Theory* **7** (1974), 355–369.
10. H. Nikaido, "Convex Structures and Economic Theory," Academic Press, New York, 1968.
11. M. L. Weitzman, Duality theory for infinite horizon convex models, *Manage. Science* **19** (1973), 783–789.

6

On Characterizing Optimality of Stochastic Competitive Processes

YAW NYARKO*

Department of Economics, Brown University, Providence, Rhode Island 02912

A condition is provided to replace the transversality condition in characterizing the optimality of competitive processes. This extends the work of W. Brock, L. Hurwicz, and M. Majumdar ("On Characterizing Optimality of Competitive Programs in Terms of Decentralizable Conditions" and "Optimal Intertemporal Allocation Mechanisms and Decentralization of Decisions," Cornell University Working Paper Nos. 333 and 369 (1985)) on providing an informationally decentralizable condition for characterizing optimality in stochastic infinite-horizon models without discounting. *Journal of Economic Literature* Classification Numbers: 026, 011.

I. INTRODUCTION

In the literature on the theory of intertemporal resource allocation it is known that optimality for infinite-horizon economies can be characterized in terms of two conditions; the first requires intertemporal profit maximization and utility maximization relative to a system of "competitive" prices, while the second is a transversality condition, which requires (for undiscounted models) that the value of the capital stock, computed at the competitive prices, be uniformly bounded over time. These results, for stochastic economies, have been proved in Zilcha [11, 12].

There are two problems with the above characterization of optimality. The first is that the transversality condition involves a limit, so one cannot verify on a period-by-period basis whether or not this condition is being attained. I shall call this the absence of *temporal decentralization*. The second problem is the absence of *informational decentralization*, in the sense of Hurwicz [8, Definition 10, p. 401]. In particular, one cannot design a

* I thank Professor Tapan Mitra for his valuable assistance. This paper is a revision of Chapter 4 of my Ph.D. thesis submitted to Cornell University, and supervised by Professor M. Majumdar. Research reported here was supported by the National Science Foundation under Grant SES 8304131 awarded to Professor M. Majumdar.

meaningful recource allocation mechanism that will ensure that the transversality condition will be met, but which is constrained so that the rules of behavior of agents at each date depend only upon the partial history at that date. One should consult [9] for futher motivation.

Brock and Majumdar [3] provide a condition to replace the transversality condition, which is both temporally and informationally decentralized. The purpose of this paper is to extend their result to handle stochastic economies; in particular to the model of intertemporal resource allocation under uncertainty discussed in [10].

The main result of this paper may be described as follows: let $(\hat{\mathbf{x}}, \hat{\mathbf{y}}, \hat{\mathbf{c}})$ be the optimal stationary process and $\{\hat{p}_t\}$ the corresponding competitive (or supporting) price process (see (3.9) and (3.10)). Then a resource allocation process $(\mathbf{x}, \mathbf{y}, \mathbf{c})$ is optimal if and only if (a) the process is competitive (see (3.7) and (3.8)) at some prices $\{p_t\}$ and (b) at each date t, $E(p_t - \hat{p}_t)(y_t - \hat{y}_t) \leqslant 0$.

The rest of this paper is organized as follows: In Section II some preliminary notation is introduced and in Section III the model is formally presented. The main results are stated in Section IV.

II. Some Preliminary Notation

R^n is the n-dimensional Euclidean space. Given any two vectors $a = (a_1, ..., a_n)$ and $b = (b_1, ..., n_n)$ in R^n we write $a \geqslant b$ if $a_i \geqslant b_i$ for each $i = 1, ..., n$; $a > b$ if $a \geqslant b$ and a is different from b; and $a \gg b$ if $a_i > b_i$ for each $i = 1, ..., n$. $R^n_+ = \{x \in R^n : x \geqslant 0\}$ and $R^n_{++} = \{x \in R^n\colon\ x \gg 0\}$. We denote by $\|\cdot\|$ the "max" norm in R^n; i.e., if $a = (a_1, ..., a_n) \in R^n$, then $\|a\| = \max\{|a_1|, ..., |a_n|\}$.

III. The Model

The framework we use is essentially the model of intertemporal resource allocation with production uncertainty, where labor is necessary for production, studied in [10, 4, 11, 12]. We shall, however, consider the version of the model that has been reduced to per capita terms. In particular the assumptions we place on the environment[1] and the technology are similar to those in [10, 4] stated in per capita terms.

[1] The assumptions we place on the environment are similar to those in [4], which is a generalization of [10]. In particular, [10] assumes that the set of states of the environment at any date is finite, the probability measure on the sequence of states, σ, is atomless, and also that the shift operator, T, is ergodic. These are used in [10] to obtain bounds on production processes. Following [4] we impose these bounds directly in assumptions (T7) and (T.8) of this paper.

III.a. *The Environment*

The environment is represented by the probability space (S, ϕ, σ), where

(i) S is the set of doubly-infinite sequences $s=(s_t)$, $-\infty<t<\infty$, with s_t representing the state of the environment at date t, which lies in some complete and separable metric space E. In particular, $S=X_{t=-\infty}^{\infty} E_t$, where for each t, $E_t=E$.

(ii) ϕ is the sigma field generated by cylinder sets in S (i.e., generated by sets of the form $X_{t=-\infty}^{\infty} B_t$, where $B_t=E$ for all but finitely many t, and where B_t belongs to the Borel sigma field on E for all t).

(iii) σ is a probability measure on ϕ, the probability distribution on the sequences of states.

The shift operator $T: S \to S$ is defined by

$$(Ts)_t = s_{t+1}. \tag{3.1}$$

We assume

(E.1) T is measure preserving.[2, 3]

Assumption (E.1) means that if A is any set in ϕ, then $\sigma(A)=\sigma(TA)$. For any integer i, $-\infty<i<\infty$, T^i is the ith iterate of T, i.e., $(T^is)_t=s_{t+i}$. Given any function f on S we define T^if by

$$T^if(s)=f(T^is). \tag{3.2}$$

Let ϕ_t denote the sigma field generated by the partial history $(..., s_{t-1}, s_t)$; ϕ_t is the sigma field generated by cylinder sets $X_{i=-\infty}^{\infty} B_i$, where $B_i=E$ for all $i>t$.

Recall that $\|\cdot\|$ denotes the norm in R^n (defined in Section II). We define for each $t=0, 1, ...,$

$$L^1(\phi_t)=\left\{f\colon S\to R^n \mid f \text{ is } \phi_t\text{-measurable and} \int \|f(s)\|\, d\sigma<\infty\right\} \tag{3.3}$$

$$L^\infty(\phi_t)=\{f\colon S\to R^n \mid f \text{ is } \phi_t\text{-measurable and essentially bounded}^4\}. \tag{3.4}$$

[2] Assumption (E.1) implies that T^{-1} is measure preserving.

[3] One consequence of the assumption that T is measure preserving is the following: Given any ϕ-measurable function f, $\int f(s)\, d\sigma=\int f(Ts)\, d\sigma$. One should consult [1] for more on measure-preserving operators.

[4] f is essentially bounded if there is a $0<C<\infty$ such that $\|f(s)\|\leqslant C$ a.s.

III.b. *The Technology*

The technology is described by the correspondence $\tau\colon R^n_+ \times S \to R^n_+$; $\tau(x, s)$ is the set of output possibilities at date 1 if the input is $x \in R^n_+$ at date 0 and the state of environment is $s \in S$. Define $B(s) = \{(x, y) \in R^n_+ \times R^n_+ \mid y \in \tau(x, s)\}$ for each $s \in S$. We impose the following assumptions on the technology:

(T.1) For all $s \in S$, $B(s)$ is closed and convex and contains (0, 0).

(T.2) For all $s \in S$, "$(x, y) \in B(s)$, $x' \geqslant x$, and $y' \leqslant y$" implies "$(x', y') \in B(s)$" [free disposal].

(T.3) For all $s \in S$, if $(0, y) \in B(s)$ then $y = 0$ [importance of inputs].

(T.4) For all $x \in R^n_+$, $\tau(x, \cdot)$ is ϕ_1-measurable (i.e., for any closed set F in R^n_+, $\{s \mid \tau(x, s) \cap F \text{ is not empty}\}$ is in ϕ_1) [measurability].

Assumption (T.4) is the requirement that production possibilities at date 1 depend (measurably) on the history of the environment up to date 1.

Define $G = \{(\alpha, \beta) \mid \alpha, T^{-1}\beta \in L^\infty(\phi_0) \text{ and } (\alpha(s), \beta(s)) \in B(s) \text{ a.s.}\}$, the set of all (stationary) production plans. Any (α, β) in G has the following interpretation: $\alpha(s)$ is the input at date 0 and $\beta(s)$ is the corresponding output at date 1 when the state is $s \in S$. We assume that there is a production plan whose net output is bounded away from zero:

(T.5) There is a $(\bar{\alpha}, \bar{\beta}) \in G$ such that $\bar{\beta}(T^{-1}s) - \bar{\alpha}(s) \geqslant \bar{c}$ a.s. for some $\bar{c} \in R^n_{++}$.

We also require the following convexity assumption on the G:

(T.6) For any (α, β) and (α', β') in G with $\alpha(s)$ different from $\alpha'(s)$ on a set D with positive measure, and for any ϕ_0-measurable random variable a, with $0 < a(s) < 1$ a.s., there exists a β'' in $L^\infty(\phi_1)$ such that $\beta''(s) \geqslant a(s)\,\beta(s) + (1 - a(s))\,\beta'(s)$ a.s. with strict inequality for any s in D, and $(a\alpha + (1 - a)\,\alpha', \beta'') \in G$ [weak strict convexity of outputs].

A *feasible process* from initial stock $y \in L^\infty(\phi_0)$ is a stochastic process $\{x_t, y_t, c_t\}_{t=0}^\infty$ which satisfies:

(a) $(x_t, y_{t+1}) \in T^tG$ for $t = 0, 1, 2, \ldots$;

(b) $y_0 = y$ and $c_t = y_t - x_t \geqslant 0$ a.s. for $t = 0, 1, 2, \ldots$,

where $T^tG = \{(\alpha_t, \beta_t)\colon \alpha_t = T^t\alpha \text{ and } \beta_t = T^t\beta \text{ for some } (\alpha, \beta) \in G\}$. (The shift operator, T, and its tth iterate, T^t, are defined in Section III.a.)

Let $P(y)$ be the set of all feasible processes from initial stock $y \in L^\infty(\phi_0)$. We shall denote a feasible process by $(\mathbf{x}, \mathbf{y}, \mathbf{c})$, where $\mathbf{x} = \{x_t\}_{t=0}^\infty$, $\mathbf{y} = \{y_t\}_{t=0}^\infty$, and $\mathbf{c} = \{c_t\}_{t=0}^\infty$; we refer to $\mathbf{x}$, $\mathbf{y}$, and $\mathbf{c}$ as the input, output, and consumption processes, respectively.

Let $G_0 = \{(\alpha, \beta) \in G \mid \beta(T^{-1}s) - \alpha(s) \geqslant 0 \text{ a.s.}\}$, the set of all feasible stationary production programs; each (α, β) in G_0 defines a stationary feasible process $(\mathbf{x}, \mathbf{y}, \mathbf{c})$, where for each $t \geqslant 0$ and $s \in S$,

$$x_t(s) = \alpha(T^t s), \qquad y_t(s) = \beta(T^{t-1}s), \qquad c_t(s) = y_t(s) - x_t(s).$$

The next two assumptions place bounds on feasible processes and stationary production programs. Recall that $\|\cdot\|$ is the norm of R^n_+ (defined in Section II).

(T.7) For each $y \in L^\infty(\phi_0)$ there is a $0 < K(y) < \infty$, such that if $(\mathbf{x}, \mathbf{y}, \mathbf{c}) \in P(y)$, then for all $t \geqslant 0$, $\|x_t(s)\| \leqslant K(y)$ a.s., $\|y_t(s)\| \leqslant K(y)$ a.s., and $\|c_t(s)\| \leqslant K(y)$ a.s.

(T.8) There is a $0 < Q < \infty$ such that for all $(\alpha, \beta) \in G_0$, $\|\alpha(s)\| \leqslant Q$ a.s. and $\|\beta(s)\| \leqslant Q$ a.s.

Remark. Note that the bound in (T.7) clearly depends on the initial stock. The bounds in (T.7) and (T.8) may be obtained by starting with the model in [10], where[1] labor is necessary for production, and then assuming that the labor supply process is bounded and finally reducing to per capita units (see [10, Theorem 3.1 and Lemma 3.2]).

III.c. *Optimality and Competitive Prices*

Feasible processes are evaluated according to the utilities generated by the corresponding consumption process. Let $u: R^n_+ \to R$ be the one-period utility function. We assume

(U.1) u is continuous on R^n_+ ;

(U.2) u is strictly increasing (i.e., $c_1 > c_2$ implies $u(c_1) > u(c_2)$);

(U.3) u is strictly concave on R^n_+.

A process $(\mathbf{x}^*, \mathbf{y}^*, \mathbf{c}^*)$ in $P(y)$ is *optimal* if for all $(\mathbf{x}, \mathbf{y}, \mathbf{c})$ in $P(y)$

$$\limsup_{N \to \infty} \sum_{t=0}^{N} [Eu(c_t) - Eu(c_t^*)] \leqslant 0, \tag{3.5}$$

where E is the expectation operator.

A program $(\hat{\alpha}, \hat{\beta}) \in G_0$ is said to be an *optimal stationary program* if

$$\sup_{(\alpha, \beta) \in G_0} \int u(\beta(T^{-1}(s)) - \alpha(s))\, d\sigma = \int u(\hat{\beta}(T^{-1}(s)) - \hat{\alpha}(s))\, d\sigma. \tag{3.6}$$

The existence of an optimal stationary program was first proved in [10].

LEMMA 3.1. (Radner). *There exists an optimal stationary program.*

Proof. See[1] [4, Corollary IV.1, p. 189]. ∎

Using the strict convexity assumption on the technology (T.6), in addition to the strict monotonicity and strict concavity assumptions on the utility function, (U.2) and (U.3), it is easy to show that the optimal stationary program, which we denote by $(\hat{x}, \hat{y})$, is unique. We define the *optimal stationary process*, $(\hat{\mathbf{x}}, \hat{\mathbf{y}}, \hat{\mathbf{c}})$, by $\hat{x}_t(s) = \hat{x}(T^t s)$, $\hat{y}_t(s) = \hat{y}(T^t s)$, and $\hat{c}_t(s) = \hat{y}_t(s) - \hat{x}_t(s)$ for all $s \in S$ and $t = 0, 1, \ldots$.

A feasible process $(\mathbf{x}, \mathbf{y}, \mathbf{c})$ is a *competitive process* if there is a sequence $\{p_t\}_{t=0}^{\infty}$ such that for each t, $p_t \in L^1(\phi_t)$, $p_t(s) > 0$ a.s., and

$$u(c_t(s)) - p_t(s)\, c_t(s) \geqslant u(c) - p_t(s)\, c \qquad \text{a.s.} \quad \text{for all } c \text{ in } R^n_+ \tag{3.7}$$

$$E[p_{t+1}(s)\, y_{t+1}(s) \mid \phi_t] - p_t(s)\, x_t(s) \geqslant E[p_{t+1}(s)\, \beta(s) \mid \phi_t] - p_t(s)\, \alpha(s) \qquad \text{a.s.} \quad \text{for all } (\alpha, \beta) \text{ in } T^t G. \tag{3.8}$$

Remark. The competitive conditions (3.7) and (3.8) are sometimes stated as

$$\int u(c_t(s))\, d\sigma - \int p_t(s)\, c_t(s)\, d\sigma \geqslant \int u(\mu(s))\, d\sigma - \int p_t(s)\, \mu(s)\, d\sigma$$
$$\text{for all } \mu \in L^{\infty}(\phi_0) \text{ with } \mu(s) \geqslant 0 \quad \text{a.s.} \tag{3.7}'$$

$$\int p_{t+1}(s)\, y_{t+1}(s)\, d\sigma - \int p_t(s)\, x_t(s)\, d\sigma \geqslant \int p_{t+1}(s)\, \beta(s)\, d\sigma - \int p_t(s)\, \alpha(s)\, d\sigma$$
$$\text{for all } (\alpha, \beta) \text{ in } T^t G. \tag{3.8}'$$

As observed by [12, Remark on p. 519], since G is defined as the set of all measurable selections of $B(s)$, it is easy to show that (3.7) and (3.8) are equivalent to (3.7)′ and (3.8)′.

The existence of competitive prices for the optimal stationary process was first shown in [10]:

LEMMA 3.2 (Radner). *There exists a system of competitive prices, $\{\hat{p}_t\}$ for optimal stationary process, $(\hat{\mathbf{x}}, \hat{\mathbf{y}}, \hat{\mathbf{c}})$. In particular, there is a $\hat{p} \in L^1(\phi_0)$ such that for all $t = 0, 1, \ldots$, $\hat{p}_t(s) = \hat{p}(T^t s) > 0$ a.s. and*

$$u(\hat{c}_t(s)) - \hat{p}_t(s)\hat{c}_t(s) \geqslant u(c) - \hat{p}_t(s)\, c \qquad \textit{a.s.} \quad \textit{for all } c \in R^n_+ \tag{3.9}$$

$$E[\hat{p}_{t+1}(s)\, \hat{y}_{t+1}(s) \mid \phi_t] - \hat{p}_t(s)\, \hat{x}_t(s) \geqslant E[\hat{p}_{t+1}(s)\, \beta(s) \mid \phi_t] - \hat{p}_t(s)\, \alpha(s) \qquad \textit{a.s.} \quad \textit{for all } (\alpha, \beta) \textit{ in } T^t G. \tag{3.10}$$

Proof. See [4, Theorem VIII.1, p. 194]. ∎

Let $(\mathbf{x}, \mathbf{y}, \mathbf{c})$ be a competitive process with supporting prices $\{p_t\}$. We define

$$v_t = (p_t - \hat{p}_t)(x_t - \hat{x}_t). \tag{3.11}$$

In characterizing the optimality of competitive processes, v_t will play an essential role. An immediate consequence of the competitive conditions (3.7)–(3.10) is the following martingale result first noted by [6]:

LEMMA 3.3. *Let* $(\mathbf{x}, \mathbf{y}, \mathbf{c})$ *be a competitive process with supporting prices* $\{p_t\}$, *and let* v_t *be as in* (3.11). *Then* v_t *is a submartingale, and, in particular,* $Ev_t \geqslant Ev_0$ *for all* $t = 0, 1, \ldots$.

Proof. From the competitive conditions (3.8) and (3.10), for each $t \geqslant 0$,

$$E[p_{t+1}(y_{t+1} - \hat{y}_{t+1}) | \phi_t] \geqslant p_t(x_t - \hat{x}_t) \quad \text{a.s.} \tag{3.12}$$

$$E[\hat{p}_{t+1}(\hat{y}_{t+1} - y_{t+1}) | \phi_t] \geqslant \hat{p}_t(\hat{x}_t - x_t) \quad \text{a.s.} \tag{3.13}$$

so by addition,

$$E[(p_{t+1} - \hat{p}_{t+1})(y_{t+1} - \hat{y}_{t+1}) | \phi_t] \geqslant (p_t - \hat{p}_t)(x_t - \hat{x}_t) \quad \text{a.s.} \tag{3.14}$$

Next, from the competitive conditions (3.7) and (3.9),

$$u(c_t) - p_t c_t \geqslant u(\hat{c}_t) - p_t \hat{c}_t \quad \text{a.s} \tag{3.15}$$

$$u(\hat{c}_t) - \hat{p}_t \hat{c}_t \geqslant u(c_t) - \hat{p}_t c_t \quad \text{a.s} \tag{3.16}$$

so by addition,

$$(p_t - \hat{p}_t)(c_t - \hat{c}_t) \leqslant 0 \quad \text{a.s.} \tag{3.17}$$

Using (3.14) and (3.17) (for $t+1$) one obtains

$$E[(p_{t+t} - \hat{p}_{t+1})(x_{t+1} - \hat{x}_{t+1}) | \phi_t] \geqslant (p_t - \hat{p}_t)(x_t - \hat{x}_t) \quad \text{a.s.} \tag{3.18}$$

from which the lemma follows immediately. ∎

We refer to $\hat{x}$ and $\hat{y}$ as the *golden rule* input and output, resp. The golden rule input, $\hat{x}$, is said to be *expansible if for some* $z \in R^n_{++}$, $(\hat{x}, T\hat{x} + z) \in G$. Clearly, if the golden rule consumption, $\hat{c}(s) = \hat{y}(T^{-1}s) - \hat{x}(s)$, satisfies $\hat{c}(s) \geqslant \Gamma$ a.s. for some Γ in R^n_{++}, then $\hat{x}$ is necessarily expansible. We now show that if the golden rule input, $\hat{x}$, is expansible, then from $\hat{x}$ a multiple d (with $d > 1$) of $\hat{x}$ can be produced.

LEMMA 3.4. *Suppose that the golden rule input,* $\hat{x}$, *is expansible; then there is a* $d > 1$ *such that* $(\hat{x}, dT\hat{x}) \in G$.

Proof. Let $I=(1,\ldots,1)\in R^n_{++}$. Since $\hat{x}$ is expansible, there is a $z=(z_1,\ldots,z_n)\in R^n_{++}$ such that $(\hat{x}, T\hat{x}+z)\in G$. Let $\bar{z}=\min\{z_1,\ldots,z_n\}$; then $z\geqslant \bar{z}I\gg 0$. Also, from (T.8) there is a $0<Q<\infty$ such that $\|T\hat{x}(s)\|\leqslant Q$ a.s.; hence $0\leqslant T\hat{x}(s)\leqslant QI$ a.s. Let $d=1+(\bar{z}/2Q)$. Then $T\hat{x}(s)+z-dT\hat{x}(s)=(1-d)\,T\hat{x}(s)+z=-(\bar{z}/2Q)\,T\hat{x}(s)+z\geqslant -(\bar{z}/2)\,I+\bar{z}I=(\bar{z}/2)\,I\gg 0$. Hence $T\hat{x}(s)+z\gg dT\hat{x}(s)$ a.s.; but then since $(\hat{x}, T\hat{x}+z)\in G$, the free disposal assumption (T.2) implies that $(\hat{x}, dT\hat{x})\in G$. ▮

The following lemma will be used in proving the results in the next section, and may be of independent interest.

LEMMA 3.5. *Suppose that the golden rule input, $\hat{x}$, is expansible. Let $0<\bar{b}<1$. Then there is an $M\geqslant 1$ such that for all $b\geqslant \bar{b}$ there is a feasible process* $(\mathbf{x},\mathbf{y},\mathbf{c})$ *with $y_0=b\hat{x}$ and $x_t(s)=\hat{x}_t(s)$ a.s. for all $t\geqslant M$.*

Proof. Fix a $0<\bar{b}<1$ and let $d>1$ be as in Lemma 3.4 above. Since $\bar{b}d^t=\bar{b}<1$ for $t=0$ and $\bar{b}d^t\to\infty$ as $t\to\infty$, there is an $M'\geqslant 0$ such that

$$\bar{b}d^t\leqslant 1 \text{ for } t\leqslant M' \qquad \text{and} \qquad \bar{b}d^t>1 \text{ for } t>M'. \tag{3.19}$$

Let $M=M'+1$. Fix any $b\geqslant\bar{b}$. We now construct a process $(\mathbf{x},\mathbf{y},\mathbf{c})$ feasible from $b\hat{x}$. Define $y_0=b\hat{x}$, $x_0=\bar{b}\hat{x}$, and $c_0=y_0-x_0=(b-\bar{b})\,\hat{x}$; for $1\leqslant t\leqslant M'$, if there exists such a t, define $y_t=x_t=\bar{b}d^t\hat{x}_t$ and $c_t=0$; and for $t>M'$, define $y_t=x_t=\hat{x}_t$ and $c_t=0$.

To show that $(\mathbf{x},\mathbf{y},\mathbf{c})$ is feasible observe that $y_0=b\hat{x}$ and $c_t=y_t-x_t\geqslant 0$ for all $t\geqslant 0$, so it remains only to show that for all $t\geqslant 0$, $(x_t, y_{t+1})\in T^tG$.

First we show that

$$(\bar{b}d^t\hat{x}_t, \bar{b}d^t\,d\hat{x}_{t+1})\in T^tG \qquad \text{for all } 0\leqslant t\leqslant M'. \tag{3.20}$$

To see this note that from (3.19), for all $0\leqslant t\leqslant M'$, $\bar{b}d^t\leqslant 1$; but from Lemma 3.4 above, $(\hat{x}_t, d\hat{x}_{t+1})\in T^tG$, and from (T.1), $(0,0)\in T^tG$, so (3.20) follows from the convexity assumption on the technology, (T.1).

We now use (3.20) to prove $(x_t, y_{t+1})\in T^tG$ for all $t=0,1,\ldots$. If $0\leqslant t<M'$, then $(x_t, y_{t+1})=(\bar{b}d^t\hat{x}_t, \bar{b}d^t d\hat{x}_{t+1})$, so from (3.20), $(x_t, y_{t+1})\in T^tG$. If $t=M'$, we obtain from (3.19) that $\bar{b}d^{t+1}>1$ so $\bar{b}d^{t+1}\hat{x}_{t+1}(s)\geqslant\hat{x}_{t+1}(s)$ a.s.; but then (3.20) and the free disposal assumption (T.2) implies that $(x_t, y_{t+1})=(\bar{b}d^t\hat{x}_t, \hat{x}_{t+1})\in T^tG$. Finally, if $t>M'$, then $(x_t, y_{t+1})=(\hat{x}_t, \hat{x}_{t+1})$, which clearly belongs to T^tG.

With $M=M'+1$, it is easy to see that $(\mathbf{x},\mathbf{y},\mathbf{c})$ satisfies all the conclusions of the lemma. ▮

IV. The Main Theorems

In this section the principal results of this paper are proved. That is, we prove that optimality can be completely characterized in terms of period-by-period conditions (see (4.1) below), thereby replacing the transversality conditions.

Theorem 4.1. *Suppose that the process* $(\mathbf{x}, \mathbf{y}, \mathbf{c})$ *is optimal from* $y_0 \gg 0$. *Then the process is competitive at some prices* $\{p_t\}$, *and*

$$E[(p_t - \hat{p}_t)(y_t - \hat{y}_t)] \leqslant 0 \qquad \text{for all } t \geqslant 0. \tag{4.1}$$

Proof. From [11, Theorem 1, p. 434] we know that the optimal process is competitive at some system of prices $\{p_t\}$.

For any $y = 0, 1, \ldots$ and $y \in L^\infty(\phi_t)$, define

$$P_t(y) = \{\{x'_i, y'_i, c'_i\}_{i=t}^\infty \mid y'_t = y, (x'_i, y'_{i+1}) \in T^i G$$

and

$$c'_i = y'_i - x'_i \geqslant 0 \qquad \text{a.s.} \qquad \text{for all } i \geqslant t\}.$$

Next, define the function $g_t(\mathbf{x}', \mathbf{y}', \mathbf{c}')$ on $P_t(y)$ by

$$g_t(\mathbf{x}', \mathbf{y}', \mathbf{c}') = \liminf_{N \to \infty} \sum_{i=t}^{N} [Eu(c'_i) - Eu(\hat{c}_i)]$$

and

$$W_t(y) = \sup\{g_t(\mathbf{x}', \mathbf{y}', \mathbf{c}') \mid (\mathbf{x}', \mathbf{y}', \mathbf{c}') \in P_t(y)\}.$$

Then from the proof of [11, Theorem 1, p. 434], since $\{x_i, y_i, c_i\}_{i=t}^\infty$ and $\{\hat{x}_i, \hat{y}_i, \hat{c}_i\}_{i=t}^\infty$ are optimal[5] from y_t and $\hat{y}_t$, resp., we obtain

$$W_t(y_t) - W_t(\hat{y}_t) \geqslant E[p_t y_t - p_t \hat{y}_t] \tag{4.2}$$

$$W_t(\hat{y}_t) - W_t(y_t) \geqslant E[\hat{p}_t \hat{y}_t - \hat{p}_t y_t]. \tag{4.3}$$

Adding (4.2) and (4.3) and rearranging results in (4.1), the theorem is proved. ∎

[5] The assertion that the optimal stationary process is optimal from its own initial stock among the set of all feasible processes is not true without the stronger convexity assumption on the technology (assumption (T.6)), as an example in [2, p. 279] with linear technology indicates. The proof of the assertion under our assumptions is as follows:

THEOREM 4.2. *Suppose that the golden rule input, $\hat{x}$, is expansible. If* $(\mathbf{x}, \mathbf{y}, \mathbf{c})$ *is a competitive process at prices* $\{p_t\}$ *and* (4.1) *holds, then the process* $(\mathbf{x}, \mathbf{y}, \mathbf{c})$ *is optimal.*

Before we prove Theorem 4.2 we will state some preliminary lemmas which may be of independent interest. In each of these lemmas $(\mathbf{x}, \mathbf{y}, \mathbf{c})$ will be a competitive process and $\{p_t\}$ will be its supporting prices.

LEMMA 4.3. *Suppose that* (4.1) *holds. Then*

$$E[(p_t - \hat{p}_t)(x_t - \hat{x}_t)] \leqslant 0 \qquad \textit{for all } t \geqslant 0. \tag{4.1$'$}$$

Proof. This follows immediately from taking expectations in (3.14) and using (4.1). ∎

LEMMA 4.4. *Suppose that* (4.1) *holds. Then there there is a* $0 < H < \infty$ *such that*

$$Ep_t x_t - Ep_t \hat{x}_t \leqslant H \qquad \textit{for all } t \geqslant 0. \tag{4.4}$$

Proof. Fix a $t = 0, 1, \ldots$. Then using (4.1)′ in Lemma 4.3 above,

$$Ep_t(x_t - \hat{x}_t) \leqslant E\hat{p}_t(x_t - \hat{x}_t) \leqslant E\hat{p}_t x_t. \tag{4.5}$$

Let $(\mathbf{x}, \mathbf{y}, \mathbf{c})$ be any feasible process from initial stock $\hat{y}_0$. Then from the competitive condition (3.10), for all $N = 0, 1, \ldots$,

$$\begin{aligned}
E \sum_{t=0}^{N} [u(c_t) - u(\hat{c}_t)] &\leqslant E \sum_{t=0}^{N} [\hat{p}_t(c_t - \hat{c}_t)] \\
&= E \sum_{t=0}^{N} [\hat{p}_t(y_t - \hat{y}_t) - \hat{p}_t(x_t - \hat{x}_t)] \\
&= E \sum_{t=0}^{N} [\hat{p}_{t+1}(y_{t+1} - \hat{y}_{t+1}) - \hat{p}_t(x_t - \hat{x}_t)] - E\hat{p}_{N+1}(y_{N+1} - \hat{y}_{N+1}) \\
&\quad + \hat{p}_0(y_0 - \hat{y}_0).
\end{aligned}$$

From the competitive condition (3.10) the summation above is nonpositive; further, from [12, Corollary 1, p. 521], we may set $\lim_{N \to \infty} E\hat{p}_{N+1}(y_{N+1} - \hat{y}_{N+1}) = 0$. Hence, if $y_0 = \hat{y}_0$, taking limits in the expression above yields

$$\limsup_{N \to \infty} E \sum_{t=0}^{N} [u(c_t) - u(\hat{c}_t)] \leqslant 0,$$

which proves the assertion.

From assumption (T.7) there is a $0 < K(y_0) < \infty$ such that $\|x_t(s)\| \leqslant K(y_0)$ a.s. (where $\|\cdot\|$ is the norm on R^n_+, defined in Section II). Hence

$$E\hat{p}_t x_t \leqslant E[\|\hat{p}_t\| \cdot \|x_t\|] \leqslant K(y_0)\, E[\|\hat{p}_t\|]. \tag{4.6}$$

However, T is measure preserving[3] and $\hat{p}_t(s) = \hat{p}(T^t s)$ with $\hat{p} \in L^1(\phi_0)$; so $E\|\hat{p}_t\| = E\|\hat{p}\| < \infty$. Thus if we define $H = K(y_0)\, E\|\hat{p}\|$ then the claim follows from (4.5) and (4.6). ▮

LEMMA 4.5. *Suppose that* (4.1) *holds and the golden rule input,* $\hat{x}$, *is expansible. Then there is a* $0 < L < \infty$ *such that for all* $N = 0, 1, 2, \ldots$,

$$\text{“}Ep_N y_N / Ep_N \hat{x}_N \geqslant \tfrac{1}{2}\text{” implies} \qquad \text{“}Ep_N y_N \leqslant L.\text{”} \tag{4.7}$$

Proof. Let $M \geqslant 1$ be the constant in Lemma 3.5 corresponding to $\bar{b} = \frac{1}{4}$. Fix an $N = 0, 1, \ldots$. Define $b_N = Ep_N y_N / 2Ep_N \hat{x}_N$ and suppose that $Ep_N y_N / Ep_N \hat{x}_N \geqslant \frac{1}{2}$; then $b_N \geqslant \frac{1}{4}$. From Lemma 3.5 there is a feasible process $(\mathbf{x}'', \mathbf{y}'', \mathbf{c}'')$ such that $y''_0 = b_N \hat{x}_0$ and $x''_M = \hat{x}_M$. We seek to "start" the process $(\mathbf{x}'', \mathbf{y}'', \mathbf{c}'')$ at date N; to this effect we define the sequence $\{x'_t, y'_t, c'_t\}_{t=N}^{\infty}$ by $x'_t(s) = x''_{t-N}(T^N s)$, $y'_t(s) = y''_{t-N}(T^N s)$, and $c'_t(s) = c''_{t-N}(T^N s)$, for all $t \geqslant N$ and $s \in S$ (where T is the shift operator and T^N is its Nth iterate, defined in Section III.a.). Observe that $y'_N = b_N \hat{x}_N$, $x'_{N+M} = \hat{x}_{N+M}$ and for all $t \geqslant N$, $(x'_t, y'_{t+1}) = (T^N x''_{t-N}, T^N y''_{t+1-N}) \in T^t G$.

From the bound placed on the technology, assumption (T.7), there is a compact set in R^n_{++} that contains $c_t(s)$ for all $t = 0, 1, \ldots$, and σ-almost every $s \in S$; since the utility function is continuous (U.1), we conclude that there is a $0 < J < \infty$ such that for all $t \geqslant 0$, $u(c_t(s)) \leqslant J$ a.s. Under assumptions (U.1) and (U.2) we may suppose without loss of generality that $u(0) = 0$ and $u(c) \geqslant 0$ for all $c \in R^n_+$. Hence

$$\sum_{t=N}^{N+M} \{Eu(c_t) - Eu(c'_t)\} \leqslant \sum_{t=N}^{N+M} Eu(c_t) \leqslant (M+1)\, J. \tag{4.8}$$

Next we use the competitive conditions (3.7) and (3.8) to obtain

$$\begin{aligned}
\sum_{t=N}^{N+M} \{Eu(c_t) - Eu(c'_t)\} &\geqslant \sum_{t=N}^{N+M} E\{p_t(c_t - c'_t)\} \qquad \text{(from (3.7))} \\
&= \sum_{t=N}^{N+M} E\{p_t(y_t - y'_t) - p_t(x_t - x'_t)\} \\
&\quad \sum_{t=N}^{N+M-1} E\{p_{t+1}(y_{t+1} - y'_{t+1}) - p_t(x_t - x'_t)\} + Ep_N(y_N - y'_N) \\
&\quad - Ep_{N+M}(x_{N+M} - x'_{N+M}) \\
&\geqslant Ep_N(y_N - y'_N) - Ep_{N+M}(x_{N+M} - x'_{N+M}) \qquad \text{(from (3.8))}.
\end{aligned} \tag{4.9}$$

Substituting $y'_N = b_N \hat{x}_N$ and $x'_{N+M} = \hat{x}_{N+M}$ in (4.9) and using the resulting expression in (4.8) yields

$$(M+1)\,J \geqslant Ep_N\, y_N - b_N\, Ep_N \hat{x}_N - Ep_{N+M}(x_{N+M} - \hat{x}_{N+M}) \quad (4.10)$$

and therefore since $b_N = Ep_N\, y_N / 2Ep_N \hat{x}_N$,

$$(M+1)\,J \geqslant \tfrac{1}{2} Ep_N\, y_N - Ep_{N+M}(x_{N+M} - \hat{x}_{N+M}). \quad (4.11)$$

But from Lemma 4.4, $Ep_{N+M}(x_{N+M} - \hat{x}_{N+M}) \leqslant H$ for some $0 < H < \infty$; using this fact in (4.11) and rearranging terms we obtain

$$Ep_N\, y_N \leqslant 2[(M+1)\,J + H]. \quad (4.12)$$

Defining $L = 2[(M+1)\,J + H]$ then concludes the proof of the lemma. ∎

We now prove Theorem 4.2.

Proof of Theorem 4.2. From [12, Theorem 1, p. 521] it suffices to show that $\sup_t Ep_t\, y_t < \infty$. Suppose, per absurdem, that this is not the case; then there is a subsequence, $\{t_k\}$, such that

$$\lim_{k \to \infty} Ep_{t_k}\, y_{t_k} = \infty. \quad (4.13)$$

We now prove the following:

Claim.

$$\limsup_{k \to \infty} Ep_{t_k} y_{t_k} / Ep_{t_k}\, \hat{x}_{t_k} \geqslant 1. \quad (4.14)$$

Proof of Claim. Fix a $t = 0, 1, \ldots$. Then

$$\frac{Ep_t\, y_t}{Ep_t \hat{x}_t} \geqslant \frac{Ep_t x_t}{Ep_t\, \hat{x}_t} = 1 + \frac{Ep_t(x_t - \hat{x}_t)}{Ep_t \hat{x}_t}. \quad (4.15)$$

From Lemma 3.3, $Ev_t \geqslant Ev_0$, where $v_t = (p_t - \hat{p}_t)(x_t - \hat{x}_t)$; a simple rearrangement of this inequality results in

$$Ep_t(x_t - \hat{x}_t) \geqslant E\hat{p}_t(x_t - \hat{x}_t) + Ev_0 \geqslant -E\hat{p}_t \hat{x}_t - |Ev_0|. \quad (4.16)$$

Since $\hat{p}_t = T^t \hat{p}$, $\hat{x}_t = T^t \hat{x}$ and T is measure preserving,[3] $E\hat{p}_t\, \hat{x}_t = E\hat{p}\hat{x}$. Substituting this in (4.16) and using the result in (4.15), we may conclude that for each $t = 0, 1, \ldots,$

$$\frac{Ep_t y_t}{Ep_t \hat{x}_t} \geqslant 1 - \frac{E\hat{p}\hat{x} + |Ev_0|}{Ep_t \hat{x}_t}. \quad (4.17)$$

Hence for each $k = 0, 1, \ldots,$

$$\frac{Ep_{t_k} y_{t_k}}{Ep_{t_k} \hat{x}_{t_k}} \geqslant 1 - \frac{E\hat{p}\hat{x} + |Ev_0|}{Ep_{t_k} \hat{x}_{t_k}}. \tag{4.17$'$}$$

Taking the lim sup of both sides of (4.17)$'$,

$$\limsup_{k \to \infty} \frac{Ep_{t_k} y_{t_k}}{Ep_{t_k} \hat{x}_{t_k}} \geqslant 1 - \frac{E\hat{p}\hat{x} + |Ev_0|}{\limsup_{k \to \infty} Ep_{t_k} \hat{x}_{t_k}}. \tag{4.18}$$

If $\limsup_{k \to \infty} Ep_{t_k} \hat{x}_{t_k} = \infty$, then (4.14) follows from (4.18). If, however, $\limsup_{k \to \infty} Ep_{t_k} \hat{x}_{t_k} < \infty$, then $Ep_{t_k} \hat{x}_{t_k}$ is uniformly bounded in k so (4.14) follows from (4.13). ∎

To complete the proof of Theorem 4.2, note that the claim implies that for infinitely many k's, $Ep_{t_k} y_{t_k}/Ep_{t_k} \hat{x}_{t_k} \geqslant \frac{1}{2}$; but then from Lemma 4.5, for all such k's, $Ep_{t_k} y_{t_k} \leqslant L$ for some $0 < L < \infty$. Hence

$$\liminf_{k \to \infty} Ep_{t_k} y_{t_k} \leqslant L < \infty, \tag{4.19}$$

which is a contradiction to (4.13).

References

1. L. Breiman, "Probability," Addison–Wesley, Reading, MA, 1968.
2. W. Brock, On existence of weakly maximal programmes in a multisector economy, *Rev. Econ. Stud.* **37** (1970), 275–280.
3. W. Brock and M. Majumdar, "On Characterizing Optimality of Competitive Programs in Terms of Decentralizable Conditions," Cornell University Working Paper No. 333, 1985.
4. R. Dana, Evaluation of development programs in a stationary stochastic economy with bounded primary resources, *in* "Mathematical Methods in Economics" (J. Los and M. Los, Eds.), North-Holland, Amsterdam, 1974.
5. S. Dasgupta and T. Mitra, Intertemporal decentralization in a multi-sector model when future utilities are discounted: A characterization, mimeo, 1985.
6. H. Folmer and M. Majumdar, On the asymptotic behavior of stochastic economic processes, *J. Math. Econ.* **5** (1978), 275–287.
7. D. Gale, On optimal development of a multisector economy, *Rev. Econ. Stud.* **34** (1967), 1–18.
8. L. Hurwicz, Optimality and informational efficiency in resource allocation processes, *in* "Studies in Resource Allocation Processes" (K. Arrow and L. Hurwicz, Eds.), Cambridge Univ. Press, London, 1977.

9. L. Hurwicz and M. Majumdar, "Optimal Intertemporal Allocation Mechanisms and Decentralization of Decisions," Cornell University Working Paper No. 369, 1985.
10. R. Radner, Optimal stationary consumption with stochastic production and resources, *J. Econom. Theory* **6** (1973), 68–90.
11. I. Zilcha, On competitive prices in a multi-spector economy with stochastic production and resources, *Rev. Econom. Stud.* **43** (1976), 431–438.
12. I. Zilcha, Transversality condition in a multi-sector economy under uncertainty, *Econometrica* **46** (1978), 515–525.

7
A Characterization of Infinite Horizon Optimality in Terms of Finite Horizon Optimality and a Critical Stock Condition

TAPAN MITRA

Cornell University

AND

DEBRAJ RAY

Indian Statistical Institute

I. INTRODUCTION

The purpose of this note is to present a characterization of infinite-horizon optimality in an aggregative stationary model. Our characterization is in terms of *finite-horizon optimality* (for every finite horizon) and a condition which essentially says that the input level on the program must be, at all times, below a *critical stock* (which is below the maximum sustainable stock).

Two aspects of this result are worth emphasizing. First, out characterization is valid for arbitrary non-convexities in the technology set; the sufficiency part of our result does not even depend on concavity of the utility function. Thus, this aspect should be of particular interest when viewed in the light of contributions to the literature on optimal intertemporal allocation under non-convexities in production. (For this literature, see the papers by Clark 1971; Skiba 1978; Majumdar and Mitra 1982, 1983; Majumdar and Nermuth 1982; Dechert and Nishimura 1983; and Mitra and Ray 1984.) For a convex technology set, and a concave utility function, finite-horizon optimality (with positive consumption in some period) can itself be characterized in terms of the Ramsey-Euler conditions or the so-called competitive conditions.

Second, our characterization of finite-horizon optimal programs which are *not* infinite-horizon optimal is in terms of a "critical stock" being exceeded by the input level on the program in *some* period. This aspect should be viewed in the spirit of the recent literature on intertemporal decentralization, where an attempt has been made to replace the usual asymptotic (transversality) condition by period-by-period verifications to signal capital overaccumulation along competitive programs. (For this literature, see the papers by Majumdar 1988; Hurwicz-Majumdar 1988; Brock-Majumdar 1988; and Dasgupta and Mitra 1988.)

II. The Model

We consider a stationary aggregative model with discounting, characterized by a *production function* $f : R_+ \to R_+$, a *discount factor* $\delta \in (0,1)$, and a *utility function* $u : R_+ \to R$. Without loss of generality we take $u(0) = 0$.

On the production function, we make the assumption:

(F) *$f(0) = 0$, f is increasing and continuous, and there is $K > 0$ such that $f(x) > x$ for $0 < x < K$, $f(x) < x$ for $x > K$.*

Below, we will invoke one or both of the following assumptions on the utility function:

(U.1) *u is continuous.*

(U.2) *u is increasing and strictly concave.*

III. Feasible Programs

Programs start from an *initial stock a*. We will suppose that initial stocks may be drawn from the interval $A \equiv (0,\alpha]$ where $\alpha < K$.

A program (**x**,**y**,**c**) is *feasible* from (some initial stock) $a > 0$ if

$$x_o \leq a \tag{1}$$

$$x_t + c_t \leq y_t \qquad \text{for } t \geq 1 \tag{2}$$

$$y_t \leq f(x_{t-1}) \qquad \text{for } t \geq 1 \tag{3}$$

IV. Optimal Programs

A feasible program (x^*,y^*,c^*) from a is *optimal* from a if it solves

$$\max \sum_{t=1}^{\infty} \delta^{t-1} u(c_t)$$

subject to (**x**,**y**,**c**) feasible from a. Under (U.1) and (F), optimal programs exist.

We will also consider the following assumption:

(P) *For each $a \in A$ and each optimal program $(\mathbf{x}^*,\mathbf{y}^*,\mathbf{c}^*)$ from a, $x_t^* > 0$ for all $t \geq 0$.*

Assumption (P) can be replaced by conditions on the utility and production functions that guarantee (P). These conditions do not necessarily require concavity of f or even u. For example, either of the following two conditions can be shown to guarantee (P), given (F) and (U.1):

(i) Unbounded steepness of u at the origin and no unbounded steepness anywhere else.

(ii) f is δ-productive near the origin and u is concave.

V. Finite Horizon Programs

Let T be an integer ≥ 1 (the *horizon*). Consider two stocks $(a,b) \geq 0$. The program $(\mathbf{x}^*,\mathbf{y}^*,\mathbf{c}^*)^T$ is *T-feasible* from a to b if

$$x_o \leq a, x_T \geq b \tag{4}$$

$$x_t + c_t \leq y_t \qquad \text{for } t = 1,\ldots,T \tag{5}$$

$$y_t \leq f(x_{t-1}) \qquad \text{for } t = 1,\ldots,T \tag{6}$$

We shall also use $(\mathbf{x},\mathbf{y},\mathbf{c})^T$ to denote the obvious restriction of a feasible program $(\mathbf{x},\mathbf{y},\mathbf{c})$ from a to its first T periods. That is, $(\mathbf{x},\mathbf{y},\mathbf{c})^T$ will be T-feasible from a to x_T.

A T-feasible program $(\mathbf{x}^*,\mathbf{y}^*,\mathbf{c}^*)$ is *T-optimal* from a to b if it solves

$$\max \sum_{t=1}^{T} \delta^{t-1} u(c_t)$$

subject to $(\mathbf{x},\mathbf{y},\mathbf{c})^T$ T-feasible from a to b.

VI. A Critical Stock

Under (F), define k^* by the condition

$$k^* \equiv \min\{s : f(s) - s \geq f(x) - x \text{ for all } x \geq 0\} \tag{7}$$

It is obvious that k^* is well defined and that $0 < k^* < K$. Recalling that $A = (0,\alpha]$, choose some number, C, such that

$$K > C > \max\{k^*,\alpha\} \tag{8}$$

C can be interpreted as a *critical stock* in the results that follow.

VII. Results

Proposition 1

Under (F), (U.1) *and* (U.2): *if* $(\mathbf{x}^*,\mathbf{y}^*,\mathbf{c}^*)$ *is optimal from a, then*
(1.a) $(\mathbf{x}^*,\mathbf{y}^*,\mathbf{c}^*)^T$ *is T-optimal from a to* x_T^* *for each* $T \geq 1$, *and*
(1.b) $x_t^* \leq C$ *for all* $t \geq 0$.

Proof Suppose $(\mathbf{x}^*,\mathbf{y}^*,\mathbf{c}^*)$ is optimal from a. Then (1.a) follows by a straightforward application of the Principle of Optimality.

To establish (1.b), note that by Mitra and Ray (1984, Proposition 4.1, Lemma 5.3 and Theorem 5.1), x_t^* converges *monotonically* to some $x^* \in [0,k^*]$. Using the definition of C, we are done. **(Q.E.D.)**

Proposition 2

Under (F), (U.1) *and* (P): *if* $(\overline{\mathbf{x}},\overline{\mathbf{y}},\overline{\mathbf{c}})$ *is feasible from a, and satisfies:*
(2.a) $(\overline{\mathbf{x}},\overline{\mathbf{y}},\overline{\mathbf{c}})^T$ *is T-optimal from a to* $\overline{x}_T$ *for each* $T \geq 1$,
(2.b) $\overline{x}_t \leq C$ *for all* $t \geq 0$,
then $(\overline{\mathbf{x}},\overline{\mathbf{y}},\overline{\mathbf{c}})$ *is optimal from a.*

Proof Suppose not. Let $(\mathbf{x}^*,\mathbf{y}^*,\mathbf{c}^*)$ be an optimal program from a. (This exists, by (F) and (U.1)).

Now, there exists $\theta > 0$ such that

$$\sum_{t=1}^{\infty} \delta^{t-1} u(c_t^*) > \sum_{t=1}^{\infty} \delta^{t-1} u(\overline{c}_t)) + \theta$$

So (again using (U.1) and (F)), there exists T' such that for all $T \geq T'$

$$\sum_{t=1}^{T} \delta^{t-1} u(c_t^*) \geq \sum_{t=1}^{\infty} \delta^{t-1} u(\overline{c}_t)) + \theta \tag{9}$$

Pick any $\hat{T} \geq T'$.

By (P), $x_{\hat{T}}^* > 0$. By (F), there is $S > \hat{T}$ such that $f^{(S-\hat{T})}(x_{\hat{T}}) \geq C$ (where $f^{(k)}$ is the k-fold composition of f with itself). Now define an S-feasible program from a to C as follows: $(\hat{\mathbf{x}},\hat{\mathbf{y}},\hat{\mathbf{c}})^S$ is described by $\hat{x}_t = x_t^*$, $t = 0,\ldots,\hat{T}$; $(\hat{y}_t,\hat{c}_t) = (y_t^*,c_t^*)$, $t = 1,\ldots,\hat{T}$; $\hat{x}_t = f^{(t-\hat{T})}(x_{\hat{T}}^*)$, $t = \hat{T}+1,\ldots,S$, $\hat{y}_t = f(\hat{x}_{t-1})$, $t = \hat{T}+1,\ldots,S$; and $\hat{c}_t = 0$, t $= \hat{T}+1,\ldots,S$.

By our convention that $u(0) = 0$,

$$\sum_{t=1}^{S}\delta^{t-1}u(\hat{c}_t)) = \sum_{t=1}^{\hat{T}}\delta^{t-1}u(c_t^*)$$

$$\geq \sum_{t=1}^{\infty}\delta^{t-1}u(\bar{c}_t)) + \theta \qquad \text{(by (9))}$$

$$\geq \sum_{t=1}^{S}\delta^{t-1}u(\bar{c}_t)) + \theta \qquad \text{(by } u(0)=0 \text{ and (2.a))} \qquad (10)$$

Now, $\hat{x}_s \geq C \geq \bar{x}_s$. So $(\hat{\mathbf{x}},\hat{\mathbf{y}},\hat{\mathbf{c}})^S$ is also feasible to $\bar{x}_s$. But then (10) contradicts (2.a). **(Q.E.D.)**

REFERENCES

BROCK, W. A., and M. MAJUMDAR (1988): "On Characterizing Optimal Competitive Programs in Terms of Decentralizable Conditions", *Journal of Economic Theory*, vol. 45, pp. 262–273.

CLARK, C. W. (1971): "Economically Optimal Policies for the Utilization of Biologically Renewable Resources", *Mathematical Biosciences*, vol. 12, pp. 245–260.

DASGUPTA, S., and T. MITRA (1988): "Characterization of Intertemporal Optimality in Terms of Decentralizable Conditions: The Discounted Case", *Journal of Economic Theory*, vol. 45, pp. 247–287.

DECHERT, W. D., and K. NISHIMURA (1983): "A Complete Characterization of Optimal Growth Paths in an Aggregated Model with a Non-Concave Production Function", *Journal of Economic Theory*, vol. 31, pp. 332–354.

HURWICZ, L., and M. MAJUMDAR (1988): "Optimal Intertemporal Allocation Mechanisms and Decentralizaton of Decisions", *Journal of Economic Theory*, vol. 45, pp. 228–261.

MAJUMDAR, M. (1988): "Decentralization in Infinite Horizon Economies: An Introduction", *Journal of Economic Theory*, vol. 45, pp. 217–227.

MAJUMDAR, M., and T. MITRA (1982): "Intertemporal Allocation with a Non-Convex Technology: The Aggregative Framework", *Journal of Economic Theory*, vol. 27, pp. 101–136.

MAJUMDAR, M., and T. MITRA (1983): "Dynamic Optimization with a Non-Convex Technology: The Case of a Linear Objective Function", *Review of Economic Studies*, vol. 50, pp. 143–151.

MAJUMDAR, M., and M. NERMUTH (1982): "Dynamic Optimization in Non-Convex Models with Irreversible Investment: Monotonicity and Turnpike Results", *Zeitschrift für Nationalökonomie*, vol. 42, pp. 339–362.

MITRA, T., and D. RAY (1984): "Dynamic Optimization on a Non-Convex Feasible Set: Some General Results for Non-Smooth Technologies", *Zeitschrift für Nationalökonomie*, vol. 44, pp. 151–175.

SKIBA, A. (1978): "Optimal Growth with a Convex-Concave Production Function", *Econometrica*, vol. 46, pp. 527–540.

8
A Necessary Condition for Decentralization and an Application to Intertemporal Allocation*

LEONID HURWICZ

Department of Economics, University of Minnesota, Minneapolis, Minnesota 55455

AND

HANS F. WEINBERGER

School of Mathematics, University of Minnesota, Minneapolis, Minnesota 55455

A necessary condition for realization by a decentralized finite-dimensional mechanism is proved. The concepts of the realization of a time sequence of allocations by a decentralized temporal process and by the particular more realistic class of decentralized evolutionary temporal processes are defined. The above condition shows that the optimal allocation in a simple model for intertemporal production and consumption can be realized by a decentralized temporal process but not by a decentralized evolutionary process which uses either a finite number of verification conditions or a finite-dimensional message space. *Journal of Economic Literature* Classification Numbers: 022, 024, 111, 113.

1. INTRODUCTION

This work is concerned with the question of whether a time sequence of resource allocations can be realized by means of a decentralized mechanism. More precisely, we are interested in welfare maximization in an infinite-horizon intertemporal (or intergenerational) economy of the type

* This work is a result of the 1983–1984 program on Mathematical Models for the Economics of Decentralized Resource Allocation at the Institute for Mathematics and its Applications. Earlier versions were presented at the IMA and at the Economics Department of the University of California at San Diego in 1984. We are grateful to the referees and the Associate Editor for numerous helpful suggestions. The first author was supported by the National Science Foundation through Grants IRI-8510042 and SES-8509547. The second author was supported by grants from the National Science Foundation and the Air Force Office of Scientific Research.

studied by Malinvaud [6], Koopmans [4], and many others since. The welfare criterion used by Malinvaud was production efficiency, and the mechanism he considered was period-by-period profit maximization, with prices treated parametrically. Malinvaud showed that, in contrast to the finite-horizon case, profit maximization does not guarantee efficiency when the horizon is infinite. An additional "transversality condition" (e.g., that the sequence of present values of the inputs converges to zero) is needed, and such a condition cannot be implemented by individual decision makers with finite lives. Therefore profit maximization does not provide a decentralized mechanism which realizes efficiency. Koopmans conjectured that it would not be possible to realize efficiency in the infinite horizon problem with any decentralized mechanism.

It is natural to inquire whether it is possible to design any decentralized mechanism, not necessarily involving perfectly competitive behavior, which would maximize some well-defined measure of welfare over some class of allocations. We shall consider an economy with a single commodity in which the rate of increase of utility is of the form $u(c_t)$, where u is a utility function and c_t is the rate of consumption in the tth time interval, while an investment of the remaining stock x_t at the beginning of this time interval results in an amount $f(x_t)$ at the end of the interval. The economy is thus described by the utility function $u(c)$ and the production function $f(x)$. As the measure of welfare we use the sum of discounted one-period utilities

$$W = \sum \delta^t u(c_t)$$

with $0 < \delta < 1$.

In order to discuss the question of whether the allocation of the optimal consumption stream $\{c_t\}$ for this model can be realized by a decentralized mechanism, we need a number of definitions. A *goal function* (or *performance function*) Q is a mapping from a set $\mathfrak{E}$ of admissible economies into an allocation space $\hat{\mathfrak{A}}$, which is a normed linear vector space.

The set $\mathfrak{E}$ is a subset of a linear vector space $\hat{\mathfrak{E}}$, and we assume that $\mathfrak{E}$ is open in the rather weak sense that its intersection with any finite-dimensional linear subspace is open. Correspondingly, we shall assume that Q is twice continuously differentiable in the sense that its restriction to the intersection of $\mathfrak{E}$ with any finite-dimensional hyperplane is twice continuously differentiable. In particular, for any e_0 in $\mathfrak{E}$ and any η in $\hat{\mathfrak{E}}$ the function $\tilde{Q}(\rho) \equiv Q(e_0 + \rho\eta)$ is a twice continuously differentiable function of ρ for small values of ρ. The derivative $\tilde{Q}'(0)$ is called the derivative (or Gâteaux derivative) of Q at e_0 in the direction η, and is denoted by $Q_e(e_0; \eta)$.

These definitions will be discussed in greater detail at the beginning of Section 2.

A *mechanism* consists of a message space[1] $\mathfrak{M}$, which is an open subset of a linear vector space $\hat{\mathfrak{M}}$; a verification function G, which is a twice continuously differentiable mapping from $\mathfrak{W} \equiv \mathfrak{E} \times \mathfrak{M}$ into a Euclidan space R^k; and an outcome function H, which is a twice continuously differentiable mapping from $\mathfrak{M}$ into the allocation space $\hat{\mathfrak{A}}$.

The differentiable mapping $G: \mathfrak{W} \to R^k$ is said to be *nondegenerate* at $w_0 \in \mathfrak{W}$ if the range (image) of its derivative $G_w(w_0; \omega)$ as ω varies is the whole space R^k.

DEFINITION 1.1. The mechanism $(\mathfrak{M}, G, H)$ is said to *realize*[2] the goal function Q if

(a) $H(m) = Q(e)$ whenever $G(e, m) = 0$;

(b) For each e in $\mathfrak{E}$ there is at least one $m \in \mathfrak{M}$ so that

(i) $G(e, m) = 0$;

(ii) G is nondegenerate at (e, m).

We now assume that the economies in $\mathfrak{E}$ are defined by two independent pieces of information, which are held by two separate agents. That is, we suppose that

$$\mathfrak{E} = \mathfrak{U} \times \mathfrak{F}.$$

In our application the two agents are the consumer and the producer, an element of $\mathfrak{U}$ represents the consumer's utility function u, and an element of $\mathfrak{F}$ is the production function f.

DEFINITION 1.2. The mechanism $(\mathfrak{M}, G, H)$ with

$$G: \mathfrak{U} \times \mathfrak{F} \times \mathfrak{M} \to R^k,$$
$$H: \mathfrak{M} \to \hat{\mathfrak{A}}$$

is said to be *decentralized* (or *privacy preserving*) if G has the form

$$G(u, f, m) = R(u, m) + S(f, m). \tag{1.1}$$

The decomposition (1.1) states that the condition $G = 0$ can be verified by an information processor (a person or a computer able to generate messages, receive responses, and compute the outcome function) which

[1] Note that the message space need not be finite-dimensional, which strengthens our impossibility result.

[2] The term *implement*, which might also be used here, is customarily reserved for mechanisms derived from a game theoretic approach.

elicits the response $R(u, m)$ to the message m from an agent who only has the information u, and the response $S(f, m)$ from a second agent who only has the information f.

When each component of R is a difference between a functional of u and a functional of m while each component of S is the difference between a functional of f and a functional of m, the mechanism consists of determining the outcome from the values of finitely many linear functionals elicited from each of the agents. If R and S do not have this form, the equilibrium state $G=0$ must usually be reached by an iterative process such as a bidding process. Such a process is called a *tâtonnement.*

We remark that it is more usual to define a decentralized mechanism by requiring one of the agents to verify a condition $g_1(u, m)=0$ and the other agent to verify another condition $g_2(f, m)=0$, where the range of g_1 lies in a Euclidean space R^{j_1} and that of g_2 lies in R^{j_2}. If we let $k=j_1+j_2$ and think of R^k as a direct product, we can set

$$G(u, f, m)=\begin{pmatrix} g_1(u, m) \\ g_2(f, m) \end{pmatrix}=\begin{pmatrix} g_1(u, m) \\ 0 \end{pmatrix}+\begin{pmatrix} 0 \\ g_2(f, m) \end{pmatrix}$$

to obtain an allocation function of the form (1.1). Conversely, if we have a G of the form (1.1), we can define the new message space $\mathfrak{M}^*=\mathfrak{M}\times R^k$ and define $g_1(u, m, m')=R(u, m)+m'$ and $g_2(f, m, m')=S(f, m)-m'$ to reduce the equation $G=0$ to the form $g_1=0$, $g_2=0$.

We now consider the case where the allocation space $\mathfrak{A}$ consists of time sequences $\{a_t\}$, where t ranges over a finite or infinite interval of nonnegative integers. In this case, the goal function Q has components Q_t, the desired allocations at times t, and if $(\mathfrak{M}, G, H)$ is a mechanism which realizes Q, then H has components H_t.

We shall call the triple $(\mathfrak{M}, G, H)$ consisting of a message space, a verification function, and an outcome function a *temporal process* if the verification function G is the direct sum of mappings G_t with finite-dimensional ranges R^{k_t}. That is, the equation $G=0$ is equivalent to the sequence of equations $G_t=0$. For the infinite-horizon problem the range of G may be infinite-dimensional, so that the concept of a temporal process is an extension of the concept of a mechanism.

We define the realization of the goal function Q by the temporal process exactly as in Definition 1.1. We use Definition 1.2 to define a decentralized temporal process.[3] This definition is, of course, equivalent to saying that each of the mappings $G_t(u, f, m)$ is of the form $R_t(u, m)+S_t(f, m)$.

[3] If the time sequence is interpreted as an intergenerational model, one thinks of different agents in different time intervals. Our concept of decentralization takes account of the fact that in each generation there are two agents who possess different private information.

We observe that "in general" a temporal process for the infinite horizon problem will have the property that the component H_0 of the outcome function cannot be determined without verifying all the conditions $G_t = 0$. That is, for some e in $\mathfrak{E}$ and for every nonnegative integer s there are two elements m and m' in $\mathfrak{M}$ such that

$$G_t(e, m) = G_t(e, m') = 0 \qquad \text{for} \quad t = 0, ..., s$$

but $H_0(m) \neq H_0(m')$.

For an intergenerational model this means that no allocation can occur at time 0 until all the future agents, even those who are as yet unborn, have verified their conditions. Such a process is thus a prescription for economic paralysis rather that a realistic model for economic behavior. For this reason we introduce a definition of a more restricted but also more reasonable class of processes.

DEFINITION 1.3. A temporal process is said to be an *evolutionary process* if it has the property that

$$G_t(e, m) = G_t(e, m') = 0 \qquad \text{for} \quad t = 0, ..., s$$

imply that

$$H_s(m) = H_s(m').$$

We note that if the goal function Q is realized by the evolutionary process $(\mathfrak{M}, G, H)$, then Q_t is realized by the mechanism

$$(\mathfrak{M}, \{G_0, G_1, ..., G_t\}, H_t).$$

Thus, it is not necessary to know the future[4] in order to make the allocation at time t.

Since we are interested in proving the impossibility of implementation with a decentralized evolutionary process, we have introduced a rather weak definition of such a mechanism. While Definition 1.3 precludes verification by future agents, it permits requiring verification by all past agents. A class of evolutionary processes which prohibit this unrealistic

[4] If one wishes to model an economy which changes in time, the space $\mathfrak{E}$ of admissible economies will also consist of time sequences, and one will need to require that G_s only depend on the components e_t with $t \leqslant s$. It is clear that one cannot realize a goal function in which Q_s depends on components e_t with $t > s$ by such a mechanism. In fact, Hurwicz and Majumdar [2] have proven the impossibility of realizing such a function by a temporal proces of a class of proposed action processes.

feature as well is obtained by letting the space $\mathfrak{M}$ consist of time sequences $\{m_t\}$ and choosing the functions G and H in the form

$$H_t(m) = h_t(m_t)$$

$$G_t(e, m) = g_t(e, m_t, h_0(m_0), h_1(m_1), ..., h_{t-1}(m_{t-1})).$$

If this process is evolutionary and if it realizes the goal function Q, then the functions $h_s(m_s)$ must have the values $Q_s(e)$. Thus if at time t the historic allocations $a_s = Q_s(e)$ for $s < t$ have been observed, the mechanism $(\mathfrak{M}_t, g_t(e, m_t, a_0, ..., a_{t-1}), h_t)$ realizes Q_t. Here $\mathfrak{M}_t$ is the range of the tth component of m as m varies over members of $\mathfrak{M}$ such that $h_s(m_s) = a_s$ for $s < t$. In particular, Q_0 is realized by the mechanism $(\mathfrak{M}_0, g_0(e, m_0), h_0)$.

In Section 2 we shall derive a rather general necessary condition which must be satisfied by a goal function Q if it is to be realized by a decentralized mechanism. The condition is that the mixed second derivative Q_{uf}, which is a bilinear form in the displacements of u and f, must vanish whenever each of these displacements satisfies a certain finite set of linear equations.

The detailed one-commodity model which we have mentioned above is formulated and analyzed in Section 3. We show that there is a unique optimal investiment and consumption stream $(\{x_t\}, \{c_t\})$, and that this goal function can be realized by an explicit decentralized temporal process which is not an evolutionary process.[5]

In Section 4 we combine the fact that if the optimal consumption stream is realized by an evolutionary process then Q_0 is realized by the mechanism $(\mathfrak{M}, G_0, H_0)$ with the results of Sections 2 and 3 to show that the infinite-horizon problem cannot be realized by a decentralized evolutionary process.[6] Theorem 4.1 gives the lower bound $k \geqslant T/2$ for the dimension of the range of the verification function G_0 in any decentralized mechanism which realizes the first allocation x_0 in the corresponding T-horizon problem. Theorem 4.2 uses the same proof to show that there is no decentralized evolutionary process which realizes x_0 in the infinite-horizon problem.

[5] Related propositions on the properties of optimal programs are found in Sobel [8] and in Majumdar and Nermuth [5]. The paper by Brock and Majumdar [1] contains independently obtained results on temporal processes. (An early version of the latter paper was circulated in 1984.)

[6] We observe that this result is only proved when the goal function is defined by maximizing the discounted total utility. In fact, V. Bala, M. Majumdar, and T. Mitra [Decentralized evolutionary realization in intertemporal economies: A possibility result, mimeo, Cornell University, July 23, 1989] have recently found that another optimality criterion for intertemporal production and consumption leads to a goal function which can be realized by a decentralized evolutionary process.

We conclude that for this model Koopmans' view is too pessimistic if one is willing to accept any temporal process, but that it is valid if one restricts oneself to evolutionary processes.

We show in Section 5 that if the verification function G satisfies a sufficiently strong solvability condition, then the dimension of the message space is at least as great as that of the range of G. In this case, the lower bound for the latter in the finite-horizon problem also gives a lower bound for the former.

Our one-commodity model is, of course, exceedingly simple. It was investigated in the expectation that if such a simple model cannot be decentralized, then, in general, a more complex one is even less likely to be capable of being decentralized. One way of tying this model into a more realistic (but still quite restricted) one is to think of $\mathfrak{E}$ as a small subeconomy of a much larger market economy and to identify the one commodity with money. In this case $f(x)-x$ could represent the return in one time period of an investment x in an optimal portfolio, while $u(c)$ would represent the utility of spending an amount c on an optimal bundle of goods and services.

2. A Necessary Condition for Decentralization

In this section we shall derive a condition which a goal function $Q(u, f)$ must satisfy if it can be realized by a decentralized mechanism. Note that we are here discussing a single mechanism rather than a temporal process.

The only information one has about the functions u and f in the mechanism problem is that they lie in certain sets $\mathfrak{U}$ and $\mathfrak{F}$ of functions. Therefore our condition will involve derivatives of mappings between spaces which may be infinite-dimensional, and we begin with a brief discussion of the differentiation process.

A *linear vector space* $\hat{\mathfrak{S}}$ is defined to be a collection of elements such as vectors, functions, or vector-valued functions with the property that if s_1 and s_2 are in $\hat{\mathfrak{S}}$ and α and β are any real numbers, then $\alpha s_1 + \beta s_2$ is also in $\hat{\mathfrak{S}}$, and the usual laws of vector addition and multiplication by a scalar hold.

We define a weak topology in any normed linear vector space $\hat{\mathfrak{S}}$ by saying that a subset $\mathfrak{B}$ of $\hat{\mathfrak{S}}$ is open if for any integer l, any collection $\eta_1, ..., \eta_l$ of elements of $\hat{\mathfrak{S}}$, and any point $b \in \mathfrak{B}$ there is an $\varepsilon > 0$ such that the linear combination

$$b + \sum_{j=1}^{l} \rho_j \eta_j$$

also lies in $\mathfrak{B}$ whenever $\rho_1^2 + \cdots + \rho_l^2 < \varepsilon^2$.

We have assumed that the allocation space is a normed linear vector space. A *normed linear vector space* is a linear vector space $\hat{\mathfrak{R}}$ in which there is a real-valued function $\|r\|$, the norm of r, with the properties that $\|r\| > 0$ for $r \neq 0$, that $\|\alpha r\| = |\alpha|\, \|r\|$ for all real α, and that $\|r+s\| \leqslant \|r\| + \|s\|$. The quantity $\|r_1 - r_2\|$ is thought of as the distance between r_1 and r_2, and convergence is defined in terms of it in the obvious way.

Any Euclidean space is, of course, a normed linear vector space.

A function $R(\rho)$ of the real variable ρ with values in the normed linear space $\hat{\mathfrak{R}}$ is said to have the *limit* r as $\rho \to \sigma$ if for each $\varepsilon > 0$ there is a $\delta > 0$ such that $|\rho - \sigma| < \delta$ implies that $\|R(\rho) - r\| < \varepsilon$. With this definition we can define continuity and differentiation of such a function.

As usual, we say that a mapping $R(s)$ from an open subset $\mathfrak{B}$ of the linear vector space $\hat{\mathfrak{S}}$ into the normed linear vector space $\hat{\mathfrak{R}}$ is *continuous* if for every $b \in \mathfrak{B}$ and every positive ε there is an open subset $\mathfrak{C}$ of $\mathfrak{B}$ which contains b and has the property that $\|R(c) - R(b)\| < \varepsilon$ for all $c \in \mathfrak{C}$.

It is easily seen that because of our definition of open set this is equivalent to saying that $R(s)$ is continuous if for every $b \in \mathfrak{B}$, every integer l, and every collection $\eta_1, \ldots, \eta_l$ of elements of $\hat{\mathfrak{S}}$ the function

$$\tilde{R}(\rho_1, \ldots, \rho_l) \equiv R\left(b + \sum_{j=0}^{l} \rho_j \eta_j\right)$$

is continuous. Similarly, we say that R is *continuously differentiable* or *twice continuously differentiable* in B if for each $b \in B$ and finite collection $\eta_j \in \hat{\mathfrak{S}}$ this function $\tilde{R}$ of the finitely many variables $\rho_1, \ldots, \rho_l$ has this property.

It is easily verified that if R is continuously differentiable, then the mapping

$$R_s(b; \eta) \equiv \left.\frac{\partial R(b + \rho_1 \eta)}{\partial \rho_1}\right]_{\rho_1 = 0}$$

is linear in η. It is continuous in both b and η in the sense defined above. The functional derivative R_s defined in this way is called the Gâteaux derivative. If the mapping R happens to have a Fréchet derivative, it will coincide with the Gâteaux derivative.

If R is twice continuously differentiable, we define the second Gâteaux derivative

$$R_{ss}(b; \eta_1, \eta_2) \equiv \left.\frac{\partial^2 R(b + \rho_1 \eta_1 + \rho_2 \eta_2)}{\partial \rho_1\, \partial \rho_2}\right]_{\rho_1 = \rho_2 = 0},$$

which is linear in η_1 and in η_2.

We shall assume that the elements u and f which define a decentralized economy lie in an open subset $\mathfrak{U}$ of a normed linear vector space $\hat{\mathfrak{U}}$ and an open subset $\mathfrak{F}$ of a normed linear vector space $\hat{\mathfrak{F}}$, respectively. We are given a twice continuously differentiable goal function $Q(u, f)$ from $\mathfrak{U} \times \mathfrak{F}$ into a normed linear vector space $\hat{\mathfrak{A}}$, the allocation space. We suppose that this goal function is realized by the decentralized mechanism $(\mathfrak{M}, G, H)$, where $\mathfrak{M}$ is an open subset of a normed linear vector space $\hat{\mathfrak{M}}$, G is a twice continuously differentiable mapping from $\mathfrak{W} = \mathfrak{U} \times \mathfrak{F} \times \mathfrak{M}$ into the Euclidean space R^k, and H is a twice continuously differentiable mapping from $\mathfrak{M}$ into $\hat{\mathfrak{A}}$. The statement that the mechanism realizes the goal function and is decentralized means that Definitions 1.1 and 1.2 are satisfied.

We shall sometimes write the points of $\mathfrak{W}$ as (u, f, m). We shall work in a neighborhood of a point $w_0 = (u_0, f_0, m_0) \in \mathfrak{W}$ at which the verification function G vanishes and is nondegenerate. (The existence of such a point for each (u_0, f_0) is assured by the assumptions contained in Definition 1.1.)

Suppose that $G(w)$ is nondegenerate at the point w_0. This means that the range of the derivative $G_w(w_0; v)$ as v varies over the space $\hat{\mathfrak{W}}$ is the whole space R^k. Therefore, for each of the coordinate unit vectors ε_j of R^k there is an element v_j such that $G_w(w_0; v_j) = \varepsilon_j$. That is,

$$[G_w(w_0; v_j)]_i = \begin{cases} 1 & \text{if } i = j \\ 0 & \text{if } i \neq j. \end{cases} \tag{2.1}$$

THEOREM 2.1. *Suppose that the mechanism $(\mathfrak{M}, G, H)$ is decentralized and that it realizes the goal function*

$$Q : \mathfrak{U} \times \mathfrak{F} \to \hat{\mathfrak{A}}.$$

Suppose further that G vanishes and is nondegenerate at the point $w_0 = (u_0, f_0, m_0)$, and that the maps G, Q, and H are twice continuously differentiable. Then Q has the following property:

For every $\zeta \in \hat{\mathfrak{U}}$ such that

$$G_u(w_0; \zeta) \equiv G_w(w_0; (\zeta, 0, 0)) = 0 \tag{2.2}$$

and every $\eta \in \hat{\mathfrak{F}}$ such that

$$G_f(w_0; \eta) \equiv G_w(w_0; (0, \eta, 0)) = 0 \tag{2.3}$$

the derivatives of Q satisfy

$$Q_u(u_0, f_0; \zeta) = Q_f(u_0, f_0; \eta) = 0 \tag{2.4}$$

and

$$Q_{uf}(u_0, f_0; \zeta, \eta) = 0. \tag{2.5}$$

Proof. Since G is nondegenerate at w_0, there exist $v_1, ..., v_k \in \hat{\mathfrak{W}}$ such that the equations (2.1) hold. Choose any $\zeta \in \hat{\mathfrak{U}}$ such that (2.2) is satisfied and any $\eta \in \hat{\mathfrak{F}}$ such that (2.3) is valid.

We note that the expression $G(w_0 + (\sigma\zeta, \rho\eta, 0) + \sum_{j=1}^{k} a_j v_j)$ is a twice continuously differentiable k-vector-valued function of the finite set of variables σ, ρ, $a_1, ..., a_k$. It vanishes when all these variables are zero, and by (2.1) its Jacobian determinant with respect to the a_j is not zero at this point. Therefore, by the implicit function theorem there exist $\varepsilon > 0$ and twice continuously differentiable functions $\alpha_1(\sigma, \rho), ..., \alpha_k(\sigma, \rho)$ such that

$$G\left(w_0 + (\sigma\zeta, \rho\eta, 0) + \sum_{j=1}^{k} \alpha_j v_j\right) = 0 \qquad \text{for} \quad \rho^2 + \sigma^2 < \varepsilon^2$$
$$\alpha_j(0, 0) = 0 \qquad \text{for} \quad j = 1, ..., k. \tag{2.6}$$

We take the partial derivative of (2.6) with respect to σ, set $\sigma = \rho = 0$, and use (2.1) and (2.2) to find that

$$\frac{\partial \alpha_j}{\partial \sigma}(0, 0) = 0 \qquad \text{for} \quad j = 1, ..., k. \tag{2.7}$$

Similarly we find that

$$\frac{\partial \alpha_j}{\partial \rho}(0, 0) = 0 \qquad \text{for} \quad j = 1, ..., k. \tag{2.8}$$

We now take the mixed second partial derivative of (2.6) with respect to σ and ρ and set $\sigma = \rho = 0$. By using the definition (1.1) of a decentralized mechanism together with (2.1), (2.7), and (2.8), we see that

$$\frac{\partial^2 \alpha_j}{\partial \rho\, \partial \sigma}(0, 0) = 0 \qquad \text{for} \quad j = 1, ..., k. \tag{2.9}$$

Since the mechanism realizes Q and since the outcome function H is independent of u and f, we have

$$Q\left(u_0 + \sigma\zeta + \sum_{j=1}^{k} \alpha_j u_j, f_0 + \rho\eta + \sum_{j=1} \alpha_j f_j\right) = H\left(m_0 + \sum_{j=1}^{k} \alpha_j m_j\right), \tag{2.10}$$

where we have written

$$w_0 = (u_0, f_0, m_0)$$

and

$$v_j = (u_j, f_j, m_j).$$

We take the derivative of both sides with respect to ρ and set $\sigma=\rho=0$. We conclude from (2.6) that

$$Q_u(u_0, f_0; \zeta)=0.$$

Similarly we conclude from (2.10) and (2.8) that $Q_f(u_0, f_0; \eta)=0$. Finally, we take the mixed derivative of (2.10) and use (2.7), (2.8), and (2.9) to obtain (2.5).

Thus the theorem is established.

Remarks. 1. In mechanism design theory the mapping G is usually an unknown of the problem. Theorem 2.1 is important because it shows that Q cannot be realized by a decentralized mechanism in which the range of the verification function has dimension k unless there are k linear homogeneous equations in ζ and k in η which together imply (2.5).

2. If one makes the stronger hypothesis that the range of the linear transformation $(G_u\ F_f)$ is all of R^k, one can choose the v_i in (2.1) so that the m-components m_i are all zero. Then the right-hand side of (2.0) is independent of the α_j and hence of σ and ρ, so that the conclusion of Theorem 2.1 follows without any assumption about the smoothness of the outcome function H. This result for the case in which all the spaces are finite-dimensional was found by Hurwicz, Reiter, and Saari [3, Theorem 4 (Theorem 4′ in the 1980 version)]. See also Saari [7].

3. A Model for Intertemporal Investment

In this section we shall derive some properties of the goal function of the following standard model for intertemporal investment. There is a production function f such that an input of the amount x of a single commodity produces an output of $f(x)$ of the same commodity after one unit of time. The commodity may be consumed or used as input for further production. There is also a utility function u such that the consumption of amount c of the commodity in one unit of time produces utility $u(c)$. The initial amount y_0 of the commodity is prescribed and the goal is to find the schedule of consumptions c_t and reinvestments x_t at the integer values of the time t which maximizes the discounted sum

$$\sum_{t=0}^{\infty} \delta^t u(c_t), \tag{3.1}$$

where δ is a given number in the interval (0, 1). The usual assumptions on the functions u and f can be written in the form

$$u \in C^2([0, X]),\ u(0)=0, \qquad u'>0,\ u''<0,$$

$$f \in C^2([0, X]),\ f(0)=0, \qquad f'>0,\ f''<0,\ \delta f'(0)>1,\ f(X)<X.$$

Here X is a number larger than any quantity of the commodity that can occur in the problem, and $C^2([0, X])$ is the usual set of functions which are twice continuously differentiable on the interval $[0, X]$. It follows from the conditions on f that there is a positive number $x^* = x^*(f, \delta)$ such that

$$\delta f'(x^*) = 1. \tag{3.2}$$

One seeks the maximum of the sum (3.1) under the constraints

$$\begin{aligned} x_{t+1} + c_{t+1} &= f(x_t) \\ x_t &\geqslant 0, \\ c_t &\geqslant 0, \\ x_0 + c_0 &= y_0, \end{aligned} \tag{3.3}$$

where the initial stock y_0 with $0 < y_0 < X$ is prescribed.

We begin with an easier problem with a finite time horizon T. This is the problem of maximizing the sum

$$\sum_{t=0}^{T} \delta^t u(c_t), \tag{3.4}$$

under the constraints (3.3) and the additional constraint that x_T is a prescribed nonnegative number[7] x_T^*. We shall always assume that x_T^* is *admissible* in the sense that

$$x_T^* < f(f(\dots f(y_0) \dots)), \tag{3.5}$$

the Tth iterate of f applied to y_0. This inequality states that x_T^* is less than the amount of the commodity one can amass by the time T by not consuming anything before this time. Because we are maximizing a continuous function over a bounded and closed finite-dimensional set, the maximum certainly exists. Because u is increasing, one finds that replacing the first equality in the constraints (3.3) by the inequality

$$x_{t+1} + c_{t+1} \leqslant f(x_t)$$

does not change the problem. In this form one is maximizing a strictly concave functional over a convex set, and it follows that there is exactly one optimizing sequence.

[7] If one does not prescribe x_T, one obtains the natural boundary condition $x_T = 0$ from the maximization. We prefer to leave the prescribed value of x_T free, because this will enable us to handle the infinite horizon problem.

By solving (3.3) for c_t and substituting in the sum (3.4), we easily derive the Euler conditions

$$-u'(c_t)+\delta f'(x_t)\,u'(c_{t+1})\begin{cases}=0 & \text{if} \quad x_t>0,\, c_t>0\\ \leqslant 0 & \text{if} \quad x_t=0\\ \geqslant 0 & \text{if} \quad c_t=0\end{cases} \qquad \text{for} \quad t=0,1,...,T-1. \tag{3.6}$$

In order to establish properties of solutions of the system (3.3), (3.6), we shall use the following lemma:

Lemma 3.1. *Let* $(\{x_t\}, \{c_t\})$ *and* $(\{\hat{x}_t\}, \{\hat{c}_t\})$ *be two solutions of the system* (3.3), (3.6) *with* $x_0<\hat{x}_0$. *Then*

$$\left.\begin{aligned} x_t&<\hat{x}_t,\\ c_t&>\hat{c}_t\end{aligned}\right. \qquad \textit{for} \quad t=1,...,T.$$

Proof. We see from the conditions $\delta f'(0)>1$ and (3.6) that since $x_t=0$ implies that $c_{t+1}=0$, it also implies that $u'(c_t)>u'(0)$. This contradicts the condition $u''<0$. Thus we always have

$$x_t>0 \qquad \text{for} \quad t<T.$$

We now prove the Lemma by induction. Clearly

$$c_0=y_0-x_0>y_0-\hat{x}_0=\hat{c}_0.$$

Suppose now that $x_t<\hat{x}_t$ and $c_t>\hat{c}_t$. Then by (3.6) and the facts that f' and u' are decreasing and $x_t>0$, $c_t>\hat{c}_t\geqslant 0$

$$u'(c_{t+1})=\frac{u'(c_t)}{\delta f'(x_t)}<\frac{u'(\hat{c}_t)}{\delta f'(\hat{x}_t)}\leqslant u'(\hat{c}_{t+1})$$

so that $c_{t+1}>\hat{c}_{t+1}$. By (3.3) and the fact that f is increasing

$$x_{t+1}=f(x_t)-c_{t+1}<f(\hat{x}_t)-\hat{c}_{t+1}=\hat{x}_{t+1}.$$

Thus the lemma is established.

It is clear from this lemma that the solution of (3.3), (3.6) with prescribed x_T is unique, and that x_t increases and c_t decreases for all t when x_T is increased. The next lemma presents another important property of solutions of this system. We recall the definition (3.2) of x^*.

LEMMA 3.2. *Let* $(\{x_t\}, \{c_t\})$ *be a solution of the system* (3.3), (3.6). *If* $x_{t-1} \leqslant x_t \leqslant x^*$, *then*

$$x_{t+1} \geqslant x_t,$$

$$c_{t+1} \geqslant c_t.$$

If $x_{t-1} \geqslant x_t \geqslant x^*$, *then*

$$x_{t+1} \leqslant x_t,$$

$$c_{t+1} \leqslant c_t.$$

Moreover, if one of the inequalities in the hypotheses is strict, the first inequality in the conclusions is strict.

Proof. Because f is strictly concave, $x_t \leqslant x^*$ implies that $\delta f'(x_t) \geqslant 1$. Then (3.6) implies that if $c_t > 0$, $u'(c_{t+1}) \leqslant u'(c_t)$ so that $c_{t+1} \geqslant c_t$. If $c_t = 0$, this is obviously also true. By (3.3) and because f is increasing, we then see that

$$x_{t+1} = f(x_t) - c_{t+1} \geqslant f(x_{t-1}) - c_t = x_t.$$

The second part of the lemma is proved in the same way.

Remark. If we define the number x_{-1} by the equation $f(x_{-1}) = y_0$, we see that the lemma is true for $t = 0$.

If we think of the sequence $\{x_t\}$ as a function of t, Lemma 3.2 states that a local minimum value of this function must lie above x^* and a local maximum value must lie below x^*. Thus any solution can have at most one local maximum or minimum, and cannot have both. In particular any value of x_t can be taken on at most twice. More exact qualitative information can be obtained from the boundary values. For example, if $y_0 < f(x^*)$ so that $x_{-1} < x^*$ and if $x_T^* < x^*$, then the sequence $\{x_t\}$ lies below x^* and has at most one maximum and no minimum.

We are now in a position to prove the existence and properties of the maximizing sequence.

THEOREM 3.1. *If the boundary value* x_T^* *is admissible in the sense that the strict inequality* (3.5) *is satisfied, the* finite-horizon *problem of maximizing the sum* (3.4) *under the constraints* (3.3) *and with* $x_T = x_T^*$ *has a unique solution* $(\{x_t\}, \{c_t\})$. *This solution satisfies the inequalities*

$$x_t > 0, c_t > 0 \qquad \textit{for} \quad t = 0, ..., T-1,$$

and is the unique solution of the constraints (3.3) *together with the equation*

$$-u'(c_t) + \delta f'(x_t)\, u'(c_{t+1}) = 0 \qquad \textit{for} \quad t = 0, 1, ..., T-1. \tag{3.7}$$

Moreover,

(a) *if* $y_0 \leqslant f(x^*)$, $x_T \leqslant x^*$ *but not both equalities hold, then* $x_t < x^*$, x_t *has no local minimum, and* c_t *is increasing;*

(b) *if* $y_0 \geqslant f(x^*)$, $x_T \geqslant x^*$ *but not both equalities hold, then* $x_t > x^*$, x_t *has no local maximum, and* c_t *is decreasing;*

(c) *if* $y_0 \leqslant f(x^*)$, $x_T \geqslant x^*$ *or* $y_0 \geqslant f(x^*)$, $x_T \leqslant x^*$, *but if not both* $y_0 = f(x^*)$ *and* $x_T = x^*$, *then* x_t *is monotone and* c_t *can have at most one maximum or minimum;*

(d) *if* $y_0 = f(x^*)$ *and* $x_T = x^*$, *then* $x_t = x^*$ *and* $c_t = f(x^*) - x^*$ *for all* t.

Proof. The existence and uniqueness of the maximizer as well as of the solution of the system (3.3), (3.6) follow from the above discussion. The property $x_t > 0$ was shown in the proof of Lemma 3.1. The monotonicity properties come directly from Lemmas 3.1 and 3.2.

The Euler conditions (3.6) will thus reduce to the Euler equation (3.7) once we prove the inequality $c_t > 0$. In order to prove this inequality, we observe that since the strict inequality (3.5) holds, a slightly larger $\hat{x}_T^*$ is also admissible. If $c_0 = 0$ in the solution of the original problem, then because y_0 is fixed, the solution $(\{\hat{x}_t\}, \{\hat{c}_t\})$ for $\hat{x}_T^*$ would have to have $\hat{c}_0 \geqslant c_0$ and hence $\hat{x}_0 \leqslant x_0$. But then Lemma 3.1 would imply that $\hat{x}_T \leqslant x_T$, so that the problem with $\hat{x}_T > x_T$ would have no solution. Since we already have proved that there is a solution, we conclude that $c_0 > 0$. We can now apply the same argument to the problem where for any $t_0 \in (0, T)$ we start with the initial stock $f(x_{t_0 - 1})$ to conclude that $c_{t_0} > 0$. This completes the proof.

We obtain a similar result for the infinite-horizon problem.

THEOREM 3.2. *The* infinite-horizon *problem of maximizing the sum* (3.1) *under the constraints* (3.3) *has a unique solution* $(\{x_t\}, \{c_t\})$. *This solution satisfies the constraints* (3.3), *the inequalities*

$$x_t > 0, c_t > 0 \qquad \textit{for} \quad t \geqslant 0,$$

and the equation (3.7). *It is also the only positive solution of the equations* (3.3) *and* (3.7) *for all nonnegative* t.

Moreover,

(a) *if* $y_0 < f(x^*)$, *then* x_t *increases to* x^* *and* c_t *decreases to* $f(x^*) - x^*$ *as* t *goes to infinity;*

(b) *if* $y_0 > f(x^*)$, *then* x_t *decreases to* x^* *and* c_t *increases to* $f(x^*) - x^*$ *as* t *goes to infinity;*

(c) *if* $y_0 = f(x^*)$, *then* $x_t = x^*$ *and* $c_t = f(x^*) - x^*$ *for all* t.

Proof. We first note that both x_t and c_t are bounded by the tth iterate $f(f(\dots f(y_0)\dots))$, which is bounded uniformly in t. Thus the difference between the sum (3.1) and the finite sum (3.4) has a bound of the form $K\delta^T$.

We denote the supremum of the sum (3.1) under the constraints (3.3) by W. Then for any positive ε there is an admissible sequence $\{\tilde{x}_t, \tilde{c}_t\}$ such that

$$\sum_{t=0}^{\infty} u(\tilde{c}_t) > W - \varepsilon.$$

Let W^T denote the maximum of the finite horizon sum (3.4) with the constraints (3.3) and the boundary condition $x_T = 0$, and let $\{x_t^T, c_t^T\}$ be the maximizing solution. Let

$$\hat{c}_t = \begin{cases} \tilde{c}_t & \text{for} \quad t < T, \\ \tilde{c}_T + \tilde{x}_T & \text{for} \quad t = T \end{cases}$$

and let $\hat{x}_T = 0$, so that $\{\hat{x}_t, \hat{c}_t\}$ is admissible for this finite horizon problem. Then

$$\sum_{t=0}^{T} \delta^t u(c_t^T) = W^T \geqslant \sum_{t=0}^{T} \delta^t u(\hat{c}_t) \geqslant \sum_{t=0}^{T} \delta^t u(\tilde{c}_t) > W - \varepsilon - K\delta^T.$$

Since this is true for arbitrary ε, we see that

$$\sum_{t=0}^{T} \delta^t u(c_t^T) \geqslant W^T - K\delta^T. \tag{3.8}$$

Since $x_T^{T+1} > 0 = x_T^T$, Lemma 3.1 shows that x_t^T is increasing and c_t^T is decreasing in T :for each fixed t. Since they are also bounded, these sequences converge to a limit sequence $\{x_t, c_t\}$ as $T \to \infty$. Because the sum in (3.8) converges uniformly in T, we may let T go to infinity to show that this limit sequence is a maximizing sequence. Its uniqueness follows as in the finite horizon case from the strict concavity of u.

Clearly if $(\{x_t\}, \{c_t\})$ is the maximizing sequence pair for the infinite-horizon problem, then for any T the restriction of this sequence to $t \leqslant T$ is the maximizer for the finite horizon problem with the value x_T specified. We observe that if the inequality (3.5) is violated, the only possible admissible sequence has $c_t = 0$ for $t < T$. Since T is arbitrary, and since a consumption stream with $c_t = 0$ for all t is clearly not optimal, we conclude that the inequality (3.5) is valid for all sufficiently large T. We therefore see from Theorem 3.1 that x_t and c_t are positive for all t, and that the sequence $\{x_t, c_t\}$ is again a solution of the equations (3.3) and (3.6). Then Lemma 3.2 shows that both x_t and c_t can have at most one maximum or minimum, so that they must eventually be monotone. Therefore they must have limiting values $\bar{x}$ and $\bar{c}$ as $t \to \infty$. Since $u' > 0$, we see from taking limits in (3.6) that $\delta f'(\bar{x}) \geqslant 1$, with the inequality possible only if $c_t = 0$ for all sufficiently large t. However, if this were the case, one could increase the sum (3.1) simply by consuming all of x_t at one of these large times. Therefore, $\bar{x} = x^*$. We then see by taking limits in the constraints (3.3) that

$$\bar{x} = x^*, \qquad \bar{c} = f(x^*) - x^*.$$

Thus we have established the limiting values in the Theorem. The monotonicity properties (a), (b), and (c) then follow immediately from Lemma 3.2.

We observe that the terms with $t \leqslant T$ of the maximizing sequence $\{x_t, c_t\}$ solve the T-horizon problem with the prescribed value x_T. If T is sufficiently large, then $c_T > 0$ and therefore x_T satisfies the admissibility inequality (3.5). We thus see from Theorem 3.1 that $x_t > 0$ and $c_t > 0$ for $t < T$. Since this is true for all sufficiently large T, the inequalities hold for all t. Therefore the conditions (3.6) reduce to (3.7).

We have thus shown that the maximizing sequence has all the properties stated in the theorem, with the possible exception of the uniqueness property. To prove this property, suppose there is a second sequence pair $(\{\hat{x}_t\}, \{\hat{c}_t\})$ with positive terms which also satisfies the conditions (3.3) and (3.7) for all nonnegative t. By the arguments given above, this sequence must converge to the same limit $(x^*, f(x^*) - x^*)$. Let $\mu_t = x_t(u, f) - \hat{x}_t$, $v_t = c_t(u, f) - \hat{c}_t$. We subtract the equations (3.3) and (3.7) for the two solutions and apply the mean value theorem to find that

$$-u''v_t + \delta f'u''v_{t+1} + \delta f''u'\mu_t = 0,$$

$$\mu_{t+1} + v_{t+1} = f'\mu_t$$

where the functions u'' and f' are evaluated at values between the two solutions. We find from these equations that

$$(\mu_{t+1} - v_{t+1})^2 = a_t^2\mu_t^2 + 2a_tb_t(-\mu_tv_t) + b_t^2v_t^2 \tag{3.9}$$

where

$$a_t = f' + \frac{2f''u'}{f'u''},$$

$$b_t = \frac{2u''}{\delta f'u''}.$$

Clearly, as t approaches infinity, b_t approaches 2 while a_t approaches a number above δ^{-1}. Hence a_t and b_t are both greater than one when t is sufficiently large. We see from Lemma 3.1 that $\mu_t \nu_t \leqslant 0$. Thus (3.9) shows that for all sufficiently large t

$$(\mu_{t+1} - \nu_{t+1})^2 \geqslant (\mu_t - \nu_t)^2. \tag{3.10}$$

Since the two original sequences have the same limits, both μ_t and ν_t and hence also their difference must approach zero. By (3.10) this can only happen if $\mu_t - \nu_t = 0$ for all sufficiently large t. Since $\mu_t \nu_t \leqslant 0$, this means that $\hat{x}_t = x_t(u, f)$ and $\hat{c}_t = c_t(u, f)$ when t is large . Now given x_{t+1} and c_{t+1}, we can find x_t from (3.7) and then c_t from (3.3). By going backward in t, we conclude that any two solutions must coincide for all nonnegative t, and the Theorem is proved.

Because Theorems 3.1 and 3.2 characterize the optimal solution $\{x_t(u, f), c_t(u, f)\}$ as the unique solution of a boundary value problem, we can realize it by a decentralized temporal process.

THEOREM 3.3. *Let*

$$\hat{\mathfrak{U}} = \{u \in C^3([0, X]) : u(0) = 0\},$$

$$\mathfrak{U} = \{u \in \hat{\mathfrak{U}} : u' > 0, u'' < 0\}$$

$$\hat{\mathfrak{F}} = \{f \in C^3([0, X]) : f(0) = 0\},$$

$$\mathfrak{F} = \{f \in \hat{\mathfrak{F}} : f' > 0, f'' < 0, \delta f'(0) > 1, f(X) < X\}.$$

The optimal solution $(\{x_t(u, f)\}, \{c_t(u, f)\})$ *of the finite or infinite horizon problem is realized by the following temporal process. The message space consists of positive sequence pairs* $(\{m_t\}, \{n_t\})$ *with* $m_t + n_t < X$, $m_0 + n_0 = y_0$, *and, if the horizon T is finite,* $m_T = x_T^*$. *The outcome function H is the defined by* $H_t(m) = (m_t, n_t)$. *Each function* G_t *has two-dimensional range and is defined by*

$$\begin{pmatrix} G_{t1} \\ G_{t2} \end{pmatrix} = \begin{pmatrix} 0 \\ -u'(n_t)/u'(n_{t+1}) \end{pmatrix} + \begin{pmatrix} m_{t+1} + n_{t+1} - f(m_t) \\ \delta f'(m_t) \end{pmatrix}.$$

We have made $\hat{\mathfrak{U}}$ and $\hat{\mathfrak{F}}$ subspaces of the space C^3 of three times continuously differentiable functions only to satisfy the requirement in the definition of mechanism that G be twice continuously differentiable in all its variables. It is clear that this particular temporal process still "works" if u and f are only required to be in $C^2([0, 1])$.

We note that each G_t only depends on a four-dimensional subspace of the message space.

Remark. We can see from the above construction that if one chooses an $x_0 < x_0(u, f)$ and solves the equations (3.3) and (3.7), there will eventually come a t such that (3.3) requires x_{t+1} to be negative. Similarly, if one takes $x_0 > x_0(u, f)$, there is a t such that (3.7) requires $u'(c_{t+1})$ to be larger than $u'(0)$, which is impossible.

If one wishes to extend Theorems 3.1, 3.2, and 3.3 to economies like the Cobb–Douglas economies in which u' and f' become infinite at 0, one needs to replace the smoothness conditions on the closed interval $[0, X]$ by the analogous conditions in the interval $(0, X]$, and the conditions $u(0)=0$ and $f(0)=0$ by $u(0+)=0$ and $f(0+)=0$, respectively. The proof of Theorem 3.1 goes through without alteration. However, the uniqueness statement of Theorem 3.2 must be altered, because there can be positive solutions of (3.3) and (3.7) which are not optimal and for which either x_t or c_t approaches zero as $t \to \infty$. Such solutions can be recognized by the fact that the product $(x_t - x^*)(c_t - f(x^*) + x^*)$ becomes negative for large t, while for the optimal solution it is nonnegative for all nonnegative t. Thus the statement of Theorem 3.2 is correct if the uniqueness statement is modified to say that the optimal solution is the only positive solution of (3.3) and (3.7) for which $(x_t - x^*)(c_t - f(x^*) + x^*)$ is nonnegative for all nonnegative t. The statement of Theorem 3.3 is correct for this case if we add an additional component $\hat{m}$ to the message and the additional component

$$G_{t3} = (x_t - x^*)(c_t + x^* - f(x^*)) - \hat{m}^2$$

in the verification function G_t.

4. The Nonexistence of a Decentralized Evolutionary Realization of the Temporal Investment Model

In this section we shall apply Theorem 2.1 to show that the infinite-horizon problem discussed in Section 3 cannot be realized by a decentralized evolutionary mechanism.

It is clear from the definition that if there were such a realization, then the mechanism $(\mathfrak{M}, G_0, H_0)$ would realize the first component $x_0(u, f)$, $c_0(u, f)$ of the optimal sequence. Since $x_0(u, f) + c_0(u, f) = y_0$, the prescribed initial stock, it is sufficient to inquire whether the specific goal function $Q(u, f) = x_0(u, f)$ can be realized by a decentralized mechanism. Thus the allocation space $\hat{\mathfrak{A}}$ for this problem is just the real line R^1. We shall consider y_0, and in the case of the finite-horizon problem also x_T^*, to be known and fixed.

We see from Definition 1.1 that if a mechanism $(\mathfrak{M}, G, H)$ realizes a goal function Q in an open subset $\mathfrak{E}$ of a linear vector space $\hat{\mathfrak{E}}$, then its restriction to the intersection of $\mathfrak{E}$ with a linear subspace of $\hat{\mathfrak{E}}$ also realizes the restriction of Q to this subset. That is, the smaller the space the easier it is to find a decentralized realization, and therefore the stronger is the statement that there is no such realization.

We shall only require $\hat{\mathfrak{U}}$ and $\hat{\mathfrak{F}}$ to contain the space C_0^∞ of all infinitely differentiable functions on the interval $[0, X]$ which vanish at 0. (See the discussion which follows the proof of Corollary 4.2).

We shall prove the following two theorems. We recall that x^* is defined in terms of δ and f by (3.2).

THEOREM 4.1. *If the decentralized mechanism* $(\mathfrak{M}, G, H)$ *realizes the time* 0 *entry* $x_0(u, f)$ *of the optimal solution* $(\{x_t\}, \{c_t\})$ *of the T-horizon problem and if* $\hat{\mathfrak{U}}$ *and* $\hat{\mathfrak{F}}$ *contain* C_0^∞, *then the dimension k of the range of G must satisfy*

$$k \geqslant \tfrac{1}{2} T.$$

THEOREM 4.2. *If* $\hat{\mathfrak{U}}$ *and* $\hat{\mathfrak{F}}$ *contain* C_0^∞, *the time zero entry* $x_0(u, f)$ *of the optimal solution* $(\{x_t\}, \{c_t\})$ *for the infinite-horizon problem cannot be realized by a decentralized mechanism.*

Theorem 4.2 is a consequence of the fact that Definition 1.2 requires k to be finite and of Theorem 4.1, which implies that no finite k will do.

From Theorem 4.2 and the definition of evolutionary process we immediately obtain the following corollary.

COROLLARY. *If* $\hat{\mathfrak{U}}$ *and* $\hat{\mathfrak{F}}$ *contain* C_0^∞, *the optimal sequence pair* $(\{x_t(u, f)\}, \{c_t(u, f)\})$ *of the infinite-horizon problem cannot be realized by a decentralized evolutionary process.*

Proof of Theorem 4.1. We write the T-horizon problem in the form

$$\begin{aligned} -u'(c_t) + \delta f'(x_t)\, u'(c_{t+1}) &= 0 \qquad \text{for} \quad t = 0, \ldots, T-1, \\ x_{t+1} + c_{t+1} - f(x_t) &= 0 \qquad \text{for} \quad t = 0, \ldots, T-1, \\ x_0 + c_0 &= y_0, \\ x_T &= x_T^*, \end{aligned} \tag{4.1}$$

where y_0 and x_T^* are prescribed.

The Jacobian matrix L of the left-hand sides of the $2T+2$ equations (4.1) with respect to the $2T+2$ unknowns (x_t, c_t) is defined by the relation

$$L\begin{pmatrix} p \\ q \end{pmatrix} = \begin{pmatrix} -u''(c_t)\, q_t + \delta f''(x_t)\, u'(c_{t+1})\, p_t + \delta f'(x_t)\, u''(c_{t+1})\, q_{t+1}]_{t=0,\ldots,T-1} \\ p_{t+1} + q_{t+1} - f'(x_t)\, p_t]_{t=0,\ldots,T-1} \\ p_0 + q_0 \\ p_T \end{pmatrix} \tag{4.2}$$

for an arbitrary $2T+2$-vector $\{p_0, p_1, ..., p_T, q_0, q_1, ..., q_T\}$. (Each of the first two rows of the right-hand side is a T-dimensional column vector.)

The following analogue of Lemma 3.2 will imply that this matrix is nonsingular.

LEMMA 4.1. *If the vector* $\{p, q\}$ *satisfies the equation*

$$L\begin{pmatrix} p \\ q \end{pmatrix} = \begin{pmatrix} r \\ s \\ 0 \\ 0 \end{pmatrix} \tag{4.3}$$

and if for some $t \leqslant T-1$ *the inequalities*

$$\begin{aligned} p_t &\geqslant 0, \\ q_t &\leqslant 0, \\ r_t &\geqslant 0, \\ s_t &\geqslant 0 \end{aligned} \tag{4.4}$$

are satisfied, then the system (4.3) *implies that also*

$$\begin{aligned} p_{t+1} &\geqslant 0, \\ q_{t+1} &\leqslant 0. \end{aligned} \tag{4.5}$$

Moreover, if one of the inequalities (4.4) *is strict, then the first inequality in* (4.5) *is strict.*

Proof. We recall that the first derivatives of f and u are positive and the second derivatives are negative. If we solve the first equation in (4.3) for q_{t+1} and use (4.4), we find the second inequality in (4.5). The second equation in (4.3) then leads to the first part of (4.5). If one of the inequalities in this chain is strict, we wind up with a stict inequality.

COROLLARY. *The matrix L defined by* (4.2) *is nonsingular.*

Proof. Lemma 4.1 applied to $(\{p_t\}, \{q_t\})$ and to $(\{-p_t\}, \{-q_t\})$, and the last two equations in (4.3) immediately show that when $r=s=0$, the only solution of (4.3) is $p=q=0$.

In view of Theorem 2.1, we can prove Theorem 4.1 by showing that if $T>2k$, then for any verification function G with k-dimensional range and one point w_0 at which (b.ii) of Definition 1.1 is satisfied one can find $\zeta \in \hat{\mathfrak{U}}$ and $\eta \in \hat{\mathfrak{F}}$ such that (2.2) and (2.3) hold but (2.5) is violated. Suppose, then, that we have such a G and such a point $w_0 = (u_0, f_0, m_0)$. We choose some $\zeta \in \hat{\mathfrak{U}}$ and some $\eta \in \hat{\mathfrak{F}}$, and we confine our attention to the two-dimensional subset $u = u_0 + \sigma\zeta$, $f = f_0 + \rho\eta$ of $\hat{\mathfrak{U}} \times \hat{\mathfrak{F}}$.

We place this (u, f) in the system (4.1). Because the Jacobian L is nonsingular, we can solve the resulting system for (x_t, c_t) as twice continuously differentiable functions of σ and ρ in some neighborhood of the origin. In order to compute the derivatives of these functions, we differentiate the system (4.1) with respect to σ and set $\sigma = \rho = 0$. We obtain the linear systems[8]

$$L\begin{pmatrix} x_f(u_0, f_0; \eta) \\ c_f(u_0, f_0; \eta) \end{pmatrix} = \begin{pmatrix} -\delta u'(c_{t+1})\,\eta'(x_t)]_{t=0,\dots,T-1} \\ \eta(x_t)]_{t=0,\dots,T-1} \\ 0 \\ 0 \end{pmatrix}. \tag{4.6}$$

and

$$L\begin{pmatrix} x_u(u_0, f_0; \zeta) \\ c_u(u_0, f_0; \zeta) \end{pmatrix} = \begin{pmatrix} \zeta'(c_t) - \delta f'(x_t)\,\zeta'(c_{t+1})]_{t=0,\dots,T-1} \\ 0]_{t=0,\dots,T-1} \\ 0 \\ 0 \end{pmatrix}. \tag{4.7}$$

If $f_0(x^*) = y_0$, we choose an $\eta \in \hat{\mathfrak{F}}$ with $\eta(x^*) \neq 0$ and $\eta'(x^*) = 0$, and replace f_0 by $f_0 + \rho\eta$ with η so small that the new point $(u_0, f_0 + \rho\eta, m_0)$ is in $\mathfrak{U} \times \mathfrak{F} \times \mathfrak{M}$ and G_w is still nondegenerate there. Thus we shall assume without loss of generality that $f_0(x^*) \neq y_0$.

Then Theorem 3.1 shows that either the sequence $\{x_t\}$ or the sequence $\{c_t\}$ is strictly monotone, while the other sequence has at most one local maximum or minimum. Suppose first that $\{x_t\}$ is monotone so that no two of its members coincide. There may, however, be pairs of c_t which are equal. If so, choose the sequence $\{c_{t,f}(u_0, f_0; \eta)\}$ in such a way that it is equal to one at one member of each pair (t, t') where $c_t = c_{t'}$ and zero for all other t. Set $x_{t,f} = 0$ for all t, and use (4.6) to find corresponding values of $\eta(x_t)$ and $\eta'(x_t)$. Since $\hat{\mathfrak{F}}$ contains all infinitely differentiable functions which vanish at 0, one can choose an η in this space with these prescribed values. In the same way as we made $f_0(x^*) \neq y_0$ in the preceding paragraph, we can now alter w_0 slightly to keep the conditions of Definition 1.1 and make the c_t all distinct.

If the sequence $\{c_t\}$ is monotone, we can make the x_t distinct by a similar process. In this case, we prescribe the values of $x_{t,u}(u_0, f_0; \zeta)$ in a

[8] Here x_f denotes the Gâteaux partial derivative with respect to f of the optimal vector x. The t component of x_f, which is the Gâteaux derivative of x_t, will be denoted by $x_{t,f}$.

suitable fashion, determine $c_{t,u}$ from the second set of equations in (4.7) and the next-to the last equation, and then find values of $\zeta'(x_t)$ so that the first set of equations in (4.7) is satisfied. We now choose a function ζ with these values and proceed as before. Thus we shall assume without loss of generality that *the x_t are all distinct and the c_t are all distinct.*

In view of this fact, we can prescribe any values p_t of the $x_{t,u}(u_0, f_0; \zeta)$ with $p_T = 0$, find corresponding $c_{t,u}(u_0, f_0; \zeta)$ from the second set of equations and the next-to-the-last equation in (4.7), and then use the first set of equations of (4.7) and the added condition $\zeta'(c_T) = 0$ to determine the values of $\zeta'(c_t)$ for $0 \leqslant t \leqslant T$ uniquely.

In particular, for any integer $s \in [0, 2k]$ we determine the derivatives $\zeta_s'(c_t)$ such that

$$x_{t,u}(u_0, f_0; \zeta_s) = \delta_{ts}, \tag{4.8}$$

the Kronecker delta.[9] We see, in particular, that $\zeta_s'(c_t) = 0$ for $t > s + 1$. Because $\hat{\mathfrak{U}}$ contains all infinitely differentiable functions which vanish at 0, we can certainly choose a function $\zeta_s \in \hat{\mathfrak{U}}$ whose derivatives take on the prescribed values at the $T + 1$ distinct points c_t. For this function ζ_s the Gâteaux derivatives $x_{t,u}$ are given by (4.8).

In order to construct a function ζ which gives a contradiction with Theorem 2.1, we shall use a linear combination

$$\zeta = \sum_{j=0}^{2k} a_j \zeta_j$$

of these functions. Since we wish to apply Theorem 2.1, we impose the k conditions

$$G_u(u_0, f_0, m_0; \zeta) = 0. \tag{4.9}$$

These give at most k linearly independent linear homogeneous equations for the coefficients a_j. Hence there is a set S of at most k (and possibly 0) of the integers in $[0, 2k]$ with the property that given any values of p_t with $t \leqslant 2k$ and $t \notin S$, one can set $a_t = p_t$ for $t \notin S$ and solve for the corresponding a_t with $t \in S$. We see from (4.8) and the form of ζ that $x_{t,u}(u_0, f_0; \zeta) = a_t$. Thus we have shown that given any vector p_t, there is a linear combination ζ of the ζ_j which satisfies (4.9) and such that $x_{t,u}(u_0, f_0; \zeta) = p_t$ for $t \notin S$.

We now choose some functions $\eta_s \in \hat{\mathfrak{F}}$ with the properties that

$$\left.\begin{aligned} \eta_s(x_t) = \eta_s'(x_t) = 0, \\ \eta_s''(x_t) = \delta_{ts} \end{aligned}\right\} \quad \text{for} \quad t = 0, \ldots, T, \; s = 0, \ldots, 2k, \tag{4.10}$$

[9] That is, δ_{ts} has the value 1 when $t = s$ and is 0 otherwise.

and let η be a linear combination

$$\eta = \sum_{s=0}^{2k} \alpha_s \eta_s.$$

We see that $\eta(x_t) = \eta'(x_t) = 0$ for all $t \leqslant T$, so that (4.6) implies that

$$x_{t,f}(f_0, u_0; \eta) = 0 \qquad \text{for all } t. \tag{4.11}$$

We now impose the conditions

$$\begin{aligned} &\alpha_s = 0 \qquad \text{for} \quad s \in S, \\ &G_f(f_0, u_0, m_0; \eta) = 0. \end{aligned} \tag{4.12}$$

Because there are at most k points in S, these constitute at most $2k$ linearly independent linear homogeneous equations in the coefficients α_s. In the same way as we constructed S, we now obtain a (possibly empty) set $S' \supset S$ of at most $2k$ points in the interval $[0, 2k]$ such that for any prescribed values of the α_s for $s \leqslant 2k$ and $s \notin S'$ there is a linear combination η of the $2k$ functions η_s which satisfies (4.12). Since S' contains at most $2k$ points, there is at least one $s_0 \notin S'$ with $s_0 \leqslant 2k < T$. We choose

$$\alpha_s = \delta_{ss_0} \qquad \text{for} \quad s \notin S'$$

and construct the corresponding η. As above, we construct a linear combination ζ of the ζ_j which satisfies (4.9) and the conditions

$$x_{t,u}(u_0, f_0; \zeta) = \delta_{ts_0} \qquad \text{for} \quad t \notin S..$$

Since $\eta''(x_t) = 0$ for $t \in S$, we see that

$$\eta''(x_t)\, x_{t,u}(u_0, f_0; \zeta) = \delta_{ts_0} \tag{4.13}$$

for all $t \leqslant T$.

We now take the mixed partial derivative of (4.1) with respect to σ and ρ, and set $\sigma = \rho = 0$. In view of (4.11), we find that

$$L \begin{pmatrix} x_{uf}(u_0, f_0; \zeta, \eta) \\ c_{uf}(u_0, f_0; \zeta, \eta) \end{pmatrix} = \begin{pmatrix} -\delta u'(c_{t+1})\, \eta''(x_t)\, x_{t,u}]_{t=0,\dots,T-1} \\ 0]_{t=0,\dots,T-1} \\ 0 \\ 0 \end{pmatrix}. \tag{4.14}$$

The property (4.13) shows that the vector in the top line on the right is zero except at the point s_0, where it is negative. The last statement of Lemma 4.1 then shows that $x_{0,uf}(u_0, f_0; \zeta, \eta) = 0$ implies that

$x_{T,uf}(u_0, f_0; \eta) < 0$, contrary to the last equation in (4.14). We conclude that $x_{0,uf}(u_0, f_0; \zeta, \eta) \neq 0$, so that the equation (2.5) with $Q = x_0$ is violated, while by (4.9) and (4.12) the conditions (2.2) and (2.3) are satisfied. Thus Theorem 2.1 shows that there cannot be a decentralized mechanism with $2k < T$, and Theorem 4.1 is established.

Proof of Theorem 4.2. In order to extend the above proof to the infinite-horizon problem, we delete the last row and set $T = \infty$ in the definition (4.2) of L, so that the first two lines on the right are infinite-dimensional vectors. We shall show in the Appendix that the infinite-dimensional matrix L is invertible, and that consequently the optimal sequence pair $(\{x_t(u, f)\}, \{c_t(u, f)\})$ of the infinite-horizon problem is twice continuously differentiable. In particular, then, the goal function $x_0(u, f)$ is twice continuously differentiable. We proceed to copy the above proof of Theorem 4.1.

As before, we can introduce a small change in f_0 if necessary to make $f_0(x^*) \neq y_0$. Then by Theorem 3.2 the sequences $\{x_t\}$ and $\{c_t\}$ are strictly monotone, so that their elements are distinct.

There are now infinitely many conditions on η_s in (4.10). However, since the x_t are strictly motone, we can find an infinitely differentiable function which satisfies all these conditions by making $\eta_s = 0$ on the interval with end points x_{s+1} and x^* and satisfying the finite set of remaining conditions outside this interval, together with the condition $\eta_s(0) = 0$. Similarly, we can find an infinitely differentiable ζ_s for which (4.8) is valid by making $\zeta_s = 0$ on the interval with end points c_{s+2} and $F(x^*) - x^*$ and satisfying finitely many conditions outside this interval.

Once this is done, the proof goes through exactly as before, and Theorem 4.2 is established.

The corollary of Theorem 4.2 follows immediately from the theorem and the earlier remark that if there is a decentralized evolutionary process which realizes the optimal sequence pair, then the decentralized mechanism $(\mathfrak{M}, G_0, H_0)$ realizes x_0.

We note that we have used the fact that $\hat{\mathfrak{U}}$ and $\hat{\mathfrak{F}}$ contain all infinitely differentiable functions which vanish at 0 only to make sure that we can find ζ_s and η_s for which ζ_s', η_s, η_s', and η_s'' take on prescribed values. In the *finite-horizon* problem there are only finitely many such conditions. Therefore, it is sufficient to assume that $\hat{\mathfrak{U}}$ and $\hat{\mathfrak{F}}$ contain all the polynomials which vanish at 0, or, more generally, any Chebyshev system.

In the *infinite-horizon* problem the requirement that $\zeta_s'(c_t) = 0$ for $t > s + 1$ means that all derivatives of ζ_s vanish at the limit point $f(x^*) - x^*$ of the c_t. If the space $\hat{\mathfrak{U}}$ contains only analytic functions, the only function which meets these requirements and vanishes at 0 is $\zeta_s \equiv 0$, for which the equation (4.8) is violated at $t = s$, and the proof of Theorem 4.2 fails. A similar argument using the requirements (4.10) for the function η_s shows that the proof also fails when $\hat{\mathfrak{F}}$ contains only analytic functions.

We do not know whether the conclusion of Theorem 4.2 is still valid when one of the spaces contains only analytic functions.

However, we note that if one of the spaces, say $\hat{\mathfrak{U}}$, has a finite dimension l, then $x_0(u, f)$ can be realized by the following decentralized parameter transfer mechanism with $k = l + 1$. We suppose that u has the form

$$u = \sum_{j=1}^{l} \alpha_j u_j$$

where the linearly independent functions $u_j(c)$ are commonly known while the coefficients α_j are known only to the consumer. The message space is R^{l+1}, the verification function is

$$G_k(u, f, m) = \begin{cases} \alpha_k - m_k & \text{for} \quad k \leqslant l \\ x_0\left(\sum_{j=1}^{l} m_j u_j, f\right) - m_{l+1} & \text{for} \quad k = l + 1, \end{cases}$$

and the outcome function is m_{l+1}.

We also note that while the functional $x_0(u, f)$ for the infinite horizon problem cannot be realized by a decentralized mechanism, it is certainly realized by the nondecentralized mechanism $(R^1, x_0(u, f) - m, m)$. In fact, the nondecentralized evolutionary mechanism with $\mathfrak{M}$ the set of sequence pairs $(\{m_t\}, \{n_t\})$, $G_t(u, f, m) = (x_t(u, f) - m_t, c_t(u, f) - n_t)$, and H the identity map clearly realizes $Q = \{x_t(u, f), c_t(u, f)\}$.

5. The Dimension of the Message Space

We observe that the dimension of the message space is not involved in the above results. Because much of the previous literature in decentralization is focussed on this dimension, we shall show that a solvability hypothesis can force the dimension of the message space to be as large as that of the range of the verification function in the mechanism. In particular, such a condition and Theorem 4.1 give a lower bound for the dimension of the message space in the finite-horizon problem.

We begin with a condition which works when the space of economies $\mathfrak{E}$ is finite-dimensional.

Definition 5.1. Let $G: \mathfrak{E} \times \mathfrak{M} \to R^k$. The equation $G(e, m) = 0$ is said to be *globally solvable* in the open subset $\mathcal{O}$ of $\mathfrak{E}$ if for each e in $\mathcal{O}$ there is an m in $\mathfrak{M}$ such that $G(e, m) = 0$.

Proposition 5.1. *Let the space $\hat{\mathfrak{E}}$ have a finite dimension p. Suppose that G is globally solvable in an open subset $\mathcal{O}$ of $\mathfrak{E}$. Also suppose that G is* *nondegenerate at each point* (e, m) *where* $e \in \mathcal{O}$ *and* $G=0$. *Then the dimension of the message space* $\mathfrak{M}$ *is at least the dimension* k *of the range of* G.

Proof. Assume that the dimension l of $\mathfrak{M}$ is less than k. Choose a compact subset $\mathcal{K}$ of $\mathcal{O}$ with nonempty interior. The implicit function theorem states that each point of $\mathcal{K} \times \mathfrak{M}$ where $G=0$ has a neighborhood in which the $p+l$ coordinates of these points where $G=0$ can be written as smooth functions of $p+l-k$ parameters. The projection of this part of the null space on $\mathfrak{E}$ is obtained by keeping only the e-part of this representation. Thus it is a piece of a manifold of dimension $p+l-k$, which is less than p. Therefore the Lebesgue measure of the projection of each such coordinate patch is zero.

Since for any integer n the set $\mathcal{K} \times \{m \in \mathfrak{M} : n \leqslant \|m\| \leqslant n+1\}$ is compact, its intersection with the null set of G can be covered with finitely many such coordinate patches. Therefore the intersection of the null set of G with the set $\mathcal{K} \times \mathfrak{M}$ is a denumerable union of the coordinate patches. Its projection on $\mathfrak{E}$ is thus a denumerable union of sets of measure zero, so that this projection has measure zero. Because K has nonempty interior, it has positive Lebesgue measure. Thus there are points e of K for which there is no m such that $G(e, m)=0$.

We have shown that when $l<k$, G is not globally solvable in $\mathcal{O}$, which proves the proposition.

When the space $\hat{\mathfrak{E}}$ is infinite-dimensional, we need a stronger condition to prove this result. (We do not know whether the conclusion of Proposition 5.1 is true when $\hat{\mathfrak{E}}$ is infinite-dimensional and G is globally but not locally solvable.)

DEFINITION 5.2. Let $G: \mathfrak{E} \times \mathfrak{M} \to R^k$. The equation $G(e, m)=0$ is said to be *locally solvable* for e at (e_0, m_0) if for any neighborhood $\mathcal{N}_1$ of m_0 there is a neighborhood $\mathcal{N}_2$ of e_0 such that for any e in $\mathcal{N}_2$ there is an m in $\mathcal{N}_1$ which makes $G(e, m)=0$.

PROPOSITION 5.2. *Let* $(\mathfrak{M}, G, H)$ *be a mechanism as defined in Section* 1. *Suppose that there is at least one point* (e_0, m_0) *at which* G *is nondegenerate and locally solvable for* e.

Then the dimension of the message space $\mathfrak{M}$ *is at least as large as the dimension* k *of the range of* G.

Proof. Let the dimension l of $\mathfrak{M}$ be $l<k$. Let $r \leqslant l$ be the dimension of the range of $G_m(u_0, f_0, m_0; \mu)$. We replace the components of G by k linearly independent linear combinations in such a way that the projection onto the first r components of G_m is already r-dimensional and that $[G_m(u_0, f_0, m_0; \mu)]_i = 0$ for all $\mu \in \hat{\mathfrak{M}}$ when $i>r$.

Because of these special properties and because G is nondegenerate, we can choose $v_1, ..., v_k$ which satisfy (2.1) and which have the additional property that if

$$v_j = (e_j, m_j),$$

then $e_j = 0$ for $j \leqslant r$. Moreover, it is easily verified that $e_{r+1}, ..., e_k$ are linearly independent. If $r < l$, we choose $\mu_1, ..., \mu_{l-r}$ so that $m_1, ..., m_r$, $\mu_1, ..., \mu_{l-r}$ form a basis for $\hat{\mathfrak{M}}$.

Then to any set of coefficients $a_{r+1}, ..., a_k$ and any element $m \in \mathfrak{M}$ there correspond coefficients $a_1, ..., a_r, \rho_1, ..., \rho_{l-r}$ such that

$$\left(e_0 + \sum_{j=r+1}^{k} a_j e_j, m\right) = (e_0, m_0) + \sum_{j=1}^{k} a_j v_j + \left(0, \sum_{\nu=1}^{l-r} \rho_\nu \mu_\nu\right).$$

Thus the question of finding an m to make $G(e_0 + \sum a_j e_j, m)) = 0$ is reduced to solving the equation

$$G\left((e_0, m_0) + \sum_{j=1}^{k} a_j v_j + \left(0, \sum_{\nu=1}^{l-r} \rho_\nu \mu_\nu\right)\right) = 0.$$

As we saw in the proof of Theorem 2.1, this equation can be solved for the k coefficients a_j as twice continuously differentiable functions α_j of the $l-r$ variables ρ_ν, when the latter are sufficiently small. Thus to an element of the form $e_0 + \sum a_j e_j$ with the a_j sufficiently small there corresponds an m which makes $G = 0$ if and only if the $k-r$-vector $\{a_{r+1}, ..., a_k\}$ lies on an $(l-r)$-dimensional smooth surface.[10] Since $l < k$, it is easy to find a vector $\{a_j\}$ arbitrarily close to the origin for which this is not the case. Thus $l < k$ implies that $G = 0$ is not locally solvable at (e_0, m_0), which proves the proposition.

We remark that Proposition 5.2 remains valid when the dimension of the verification function G is infinite, provided the smoothness of G is interpreted as the smoothness of each of its components, and the nondegeneracy is taken to mean that for every positive integer k there are k components of G such that the mapping which consists of only these components is nondegenerate. Since the equation $G = 0$ implies that these k components of G are zero, Proposition 5.2 then shows that $l \geqslant k$ for every positive integer k, which means that the message space $\mathfrak{M}$ cannot be finite-dimensional.

APPENDIX: THE DIFFERENTIABILITY OF THE OPTIMAL SEQUENCE

In this Appendix we shall prove the following result, which is needed for the proof of Theorem 4.2.

[10] If $r = l$, this "surface" is the single point $a_{r+1} = \cdots = a_k = 0$.

THEOREM A.1. *The optimal sequence pair* $(\{x_t(u,f)\}), \{c_t(u,f)\})$ *of the infinite-horizon problem solved in Section* 3 *is twice continuously differentiable in the sense defined in Section* 2.

Proof. We recall that the optimal sequence is characterized as the unique positive solution of the conditions (3.7) and (3.3). That is,

$$P(u, f, \{x_t\}, \{c_t\}) \equiv \begin{pmatrix} \{-u'(c_t) + \delta f'(x_t)\, u'(c_{t+1})\} \\ \{x_{t+1} + c_{t+1} - f(x_t)\} \\ x_0 + c_0 - y_0 \end{pmatrix} = 0. \quad \text{(A.1)}$$

We define the Banach space[11]

$$b_0 = \{\{p_t\}, \{q_t\} : \lim_{t\to\infty} \delta^t(|p_t| + |q_t|) = 0\}$$

with the norm

$$\|\{p_t\}, \{q_t\}\| = \max_{t\geq 0} \delta^t(|p_t| + |q_t|).$$

It is easily seen that for fixed twice continuously differentiable functions u and f the transformation P takes a pair of sequences $(\{x_t\}, \{c_t\})$ with $(\{x_t - x^*(f)\}, \{c_t - f(x^*) + x^*\}) \in b_0$ into a pair of sequences in b_0 followed by a number. That is, we can think of P as a nonlinear operator on b_0.

We now write the linearization

$$L\begin{pmatrix} \{p_t\} \\ \{q_t\} \end{pmatrix} = \begin{pmatrix} \{-u_0''(c_t^0)\, q_t + \delta f_0'(x_t^0)\, u_0''(c_{t+1}^0)\, q_{t+1} + \delta f_0''(x_t^0)\, u_0'(c_{t+1}^0)\, p_t\} \\ \{p_{t+1} + q_{t+1} - f_0'(x_t^0)\, p_t\} \\ p_0 + q_0 \end{pmatrix} \quad \text{(A.2)}$$

of $P(u_0, f_0, \{x_t\}, \{c_t\})$ about the point $(u_0, f_0, \{x_t^0\}, \{c_t^0\})$, where we have introduced the abbreviations

$$x_t^0 = x_t(u_0, f_0), \qquad c_t^0 = c_t(u_0, f_0).$$

The linear transformation L again takes b_0 into b_0. We wish to show that it is a one-to-one transformation with bounded inverse defined on all of b_0.

[11] A Banach space is a normed linear vector space with the additional completeness property that every sequence which satisfies the Cauchy criterion has a limit in the space.

LEMMA A.1. *The equation*

$$L\begin{pmatrix}\{p_t\}\\ \{q_t\}\end{pmatrix}=\begin{pmatrix}\{r_t\}\\ \{s_t\}\\ 0\end{pmatrix} \tag{A.3}$$

has a unique solution $(\{p_t\},\{q_t\})\in b_0$ *for every* $(\{r_t\},\{s_t\})\in b_0$, *and the norm of the solution is bounded by a constant times the norm of* $(\{r_t\},\{s_t\})$.

Proof. We begin by defining the sequence

$$\gamma_t=\prod_{s=0}^{t-1} f_0'(x_t^0) \tag{A.4}$$

and the new variables

$$\mu_t=p_t/\gamma_t. \tag{A.5}$$

It is easily seen that $x_t^0-x^*$ approaches zero at an exponential rate as t goes to infinity, where $\delta f_0'(x^*)=1$. It follows that the sequence $\delta^t\gamma_t$ converges to a positive number as $t\to\infty$. Therefore there are positive constants A and B such that

$$A\delta^{-t}\leqslant\gamma_t\leqslant B\delta^{-t}. \tag{A.6}$$

In terms of the new variables μ_t the equation (A.3) becomes

$$\begin{aligned}-u_0''(c_t^0)\,q_t+\delta f_0'(x_t^0)\,u_0''(c_{t+1}^0)\,q_{t+1}+\delta f_0''(x_t^0)\,u_0'(c_{t+1}^0)\,\gamma_t\mu_t&=r_t,\\ \gamma_{t+1}[\mu_{t+1}-\mu_t]+q_{t+1}&=s_t,\\ p_0+q_0&=0.\end{aligned}$$

Because $\delta^t p_t$ goes to zero as $t\to\infty$, we see from (A.5) and (A.6) that μ_t goes to zero. Thus if μ_t has any positive values, it must take on a positive maximum. If such a maximum occurs at $t\geqslant 1$, then $\mu_t-\mu_{t-1}\geqslant 0$ so that by the second set of equations $q_t\leqslant s_t$, and for a similar reason $q_{t+1}\geqslant s_{t+1}$. We substitute these inequalities in the first set of equations and recall that the first derivatives of u_0 and f_0 are positive and their second derivatives are negative to conclude that the positive maximum must satisfy the inequality

$$p_t\leqslant\frac{r_t+u_0''(c_t^0)\,s_t-\delta f_0'(x_t^0)\,u_0''(c_{t+1}^0)\,s_{t+1}}{\delta f_0''(x_t^0)\,u_0'(c_{t+1}^0)\,\gamma_t}.$$

If the positive maximum occurs at $t=0$, we again have $q_{t+1}\geqslant 0$, and the last equation gives the equality $q_0 = -p_0$. By substituting in the $t=0$ equation of the first set of (A.3) we find that

$$p_0 \leqslant \frac{r_0 - \delta f_0'(x_0^0)\, u_0''(c_1^0)\, s_0}{u_0''(c_0^0) + \delta f_0''(x_0)\, u_0'(c_1^0)}.$$

The same reasoning leads to the same bounds with the inequalities reversed for a negative minimum. Because of the factor γ_t on the right and the inequalities (A.6), we see that there is a constant D such that

$$\max_t |\mu_t| \leqslant D\, \|(\{r_t\}, \{s_t\})\|.$$

It then follows from the transformation (A.5) and the inequalities (A.6) that

$$\delta^t\, |p_t| \leqslant BD\, \|(\{r_t\}, \{s_t\})\|.$$

We use the second set of equations in (A.3) and the last equation to solve for each q_t as a linear combination with bounded coefficients of p_t, p_{t-1}, and s_{t-1}. This together with the preceding inequality shows that there is a constant C such that

$$\|\{p_t\}, \{q_t\}\| \leqslant C\, \|\{r_t\}, \{s_t\}\|.$$

This inequality immediately shows that there is at most one solution of (A.3), that L has a bounded inverse on its range, and that this range is a closed linear subspace of b_0. Thus if the range of L is not the whole space, there must be a bounded linear functional which vanishes on the range. It is easily seen that any linear bounded linear functional on b_0 can be written in the form

$$l[(\{r_t\}, \{s_t\})] = \sum_{t=0}^{\infty} (m_t r_t + n_t s_t),$$

where $(\{m_t\}, \{n_t\})$ is some sequence pair with the property

$$\sum_{t=0}^{\infty} \delta^{-t}(|m_t| + |n_t|) < \infty.$$

If this linear functional vanishes on the range of L, then for any choice of $(\{p_t\}, \{q_t\}) \in b_0$ the result of applying it to $L(\{p_t\}, \{q_t\})$ must be zero. If, in particular, we successively let $p_t = \delta_{ts}$, $q_t = 0$ for $s = 1, 2, \ldots$, then let $p_t = 0$, $q_t = \delta_{ts}$, and finally set $p_t = \delta_{t0}$, $q_t = -\delta_{t0}$, we obtain the system of equations

$$\delta f_0''(x_s^0)\, u_0'(c_{s+1}^0)\, m_s + n_{s-1} - f_0'(x_s^0)\, n_s = 0,$$

$$-u_0''(c_s^0)\, m_s + \delta f_0'(x_{s-1}^0)\, u_0''(c_s^0)\, m_{s-1} + n_{s-1} = 0,$$

$$u_0''(c_0^0)\, m_0 + f_0'(x_0^0) n_0 = 0.$$

The reasoning used above shows that the sequence n_s/γ_{s+1} vanishes at infinity and cannot attain a positive maximum or a negative minimum. Therefore all the n_s are zero, and it follows from the first set of equations and the last equation that all the m_s are also zero.

Thus there is no nontrivial bounded linear functional which vanishes on the range of L. We conclude that this range is the whole space b_0, and Lemma A.1 is established.

We return to the proof of Theorem A.1. Lemma A.1 shows that the operator L has a bounded inverse operator L^{-1}. We now let $u = u_0 + \sum \sigma_j \zeta_j$ and $f = f_0 + \sum \rho_k \eta_k$. As in the standard proof of the implicit function theorem, we use L^{-1} to rewrite the equation (A.1) in the form

$$\begin{pmatrix} x_t - x_t^0 \\ c_t - c_t^0 \end{pmatrix} = -L^{-1}[P(u_0, f_0; \{x_t\}, \{c_t\}) - L(\{x_t - x_t^0\}, \{c_t - c_t^0\}) - P(u_0, f_0; \{x_t^0\}, \{c_t^0\})] - L^{-1}\begin{pmatrix} \sum \sigma_j \zeta_j(c_t^0) + \delta f_0'(x_t^0) \sum \sigma_j \zeta_j(c_{t+1}^0) \\ + \delta \sum \rho_k \eta_k(x_t^0)[u_0'(c_{t+1}^0) + \sum \sigma_j \zeta_j(c_{t+1}^0)] \\ -\sum \rho_k \eta_k(x_t^0) \\ 0 \end{pmatrix}, \tag{A.7}$$

The norm of the first term on the right is clearly of order $\|(\{x_t - x_t^0\}, \{c_t - c_t^0\})\|^2$, while that of the other term is of order $|\sigma| + |\rho|$, the sum of the Euclidean norms. It follows from (A.7) that $\|(\{x_t - x_t^0\}, \{c_t - c_t^0\})\|$ is of order $|\sigma| + |\rho|$ when this quantity is small. Therefore the left-hand side is equal to the second term on the right, which is a polynomial in σ and ρ, plus a term of order $(|\sigma| + |\rho|)^2$.

If we substitute this fact back into the equation (A.7), we can write the right-hand side as a quadratic polynomial in σ and ρ plus a term of order $(|\sigma| + |\rho|)^3$. Thus we see that the function $(\{x_t(u_0 + \sum \sigma_j \zeta_j, f_0 + \sum \rho_k \eta_k)\}, \{c_t(u_0 + \sum \sigma_j \zeta_j, f_0 + \sum \rho_k \eta_k)\})$ is twice continuously differentiable at $\rho = \sigma = 0$. Since we can replace (u_0, f_0) by any nearby point of the form $(u_0 + \sum \sigma_j \zeta_j, f_0 + \sum \rho_k \eta_k)$ in the above argument, we find that the function $(\{x_t(u_0 + \sum \sigma_j \zeta_j, f_0 + \sum \rho_k \eta_k)\}, \{c_t(u_0 + \sum \sigma_j \zeta_j, f_0 + \sum \rho_k \eta_k)\})$ is a twice continuously differentiable function of σ and ρ. According to our definition, then, the function $(\{x_t(u, f)\}, \{c_t(u, f)\})$ is twice continuously differentiable.

REFERENCES

1. W. A. BROCK AND M. MAJUMDAR, On characterizing optimal competitive programs in terms of decentralizable conditions, *J. Econ. Theory* **45** (1988), 262–273.
2. L. HURWICZ AND M. MAJUMDAR, Optimal intertemporal mechanisms and decentralization of decisions, *J. Econ. Theory* **45** (1988), 228–261.
3. L. HURWICZ, S. REITER, AND D. SAARI, On constructing mechanisms with message spaces of minimal dimensions for smooth performance functions, preprint, Conference Seminar on Decentralization, University of Minnesota, April, 1978; revised preprint, March, 1980.
4. T. C. KOOPMANS, "Three Essays on the State of Economic Science," McGraw–Hill, New York, 1957.
5. M. MAJUMDAR AND M. NERMUTH, Dynamic optimization in nonconvex models with irreversible investment: Monotonicity and turnpike results, *Z. Nationalökon. J. Econ.* **42** (1982), 339–362.
6. E. MALINVAUD, Capital accumulations and efficient allocation of resources, *Econometrica* **21** (1953), 233–268.
7. D. SAARI, A method for constructing message systems for smooth performance functions, *J. Econ. Theory* **33** (1978), 249–274.
8. J. SOBEL, Fair allocations of a renewable resource, *J. Econ. Theory* **21** (1979), 235–248.

9

Decentralized Evolutionary Mechanisms for Intertemporal Economies:

A Possibility Result

By

Venkatesh Bala, Mukul Majumdar, and **Tapan Mitra,** Ithaca, New York, USA*

We consider a stationary, infinite horizon aggregative model with one consumer and one producer living in each period. A decentralized intertemporal mechanism, satisfying the following "evolutionary" property, is constructed: if the current period's producer and consumer verify their equilibrium conditions, then the allocation is actually executed, without further verification by future agents. The mechanism is based on the idea of continual planning revision. It is shown that the outcome is an intertemporally efficient allocation which maximizes the long run average of one period utilities from consumption.

1. Introduction

From Adam Smith onwards, a long line of economists "... have sought to show that a decentralized economy motivated by self-interest and guided by price signals would be compatible with a coherent disposition of economic resources that could be regarded, in a well-defined sense, as superior to a large class of possible alternative dispositions." (Arrow and Hahn, 1971, p. vii).

In a static world with finitely many commodities and consumers, economists have largely succeeded in establishing that a competitive

* We would like to thank L. Hurwicz, E. Malinvaud, and R. Radner for valuable discussions, and two referees for helpful comments. Research on this project was partially supported by a National Science Foundation Grant.

equilibrium in a convex economy without externalities (that is, in a "classical environment") achieves a Pareto optimal outcome. However, once we introduce time and look at an economy over an infinite horizon, the welfare theorems break down, even in a classical environment. Thus, Malinvaud (1953) provides an example of an infinite horizon economy where even though producers are maximizing intertemporal profits in every period the (inefficient) outcome is one with zero consumption in all periods! Going beyond efficiency, Samuelson's (1958) overlapping generations model shows that a competitive equilibrium need not be Pareto optimal once an infinity of commodities and agents are admitted.

These examples suggest that a decentralized resource allocation mechanism using a "competitive" price system to guide allocation decisions can lead to sub-optimal outcomes in an infinite horizon classical economy. One is then led to ask: is it possible to design any (not necessarily price-guided) decentralized infinite-horizon mechanism which realizes efficient/optimal outcomes for an interesting class of environments?

Noting Malinvaud's result that in addition to intertemporal profit maximization, a "transversality" condition (that is, the present value of the input sequence converges to zero) guarantees efficiency, Koopmans (1957) conjectured that the answer to the above question is "No," since no finitely-lived agent can actually verify such a requirement.

Koopmans' conjecture, strictly speaking, turns out to be untrue, as Dasgupta and Mitra (1988a, 1988b), Brock and Majumdar (1988) and Hurwicz and Weinberger (1990) have demonstrated in the context of alternative models. For example, condition (S) in Majumdar (1988), along with the competitive conditions for intertemporal economies, provides a rule by which optimal plans can be attained via period-by-period verifications by agents living in each period. As a consequence it is possible, at least in principle, to obtain intertemporal optimality through a decentralized system.

However, a new issue arises here. Paraphrasing from Hurwicz (1986, p. 244), we can imagine that allocations in a static Arrow-Debreu economy are made in the following way: the economic agents are presented with a proposed message (prices and allocation); if all agents accept it as an equilibrium (that is, say "yes") then the allocation is carried out. If someone says "no," a new message must be proposed and verified by the agents, the process continuing until an equilibrium message is found, to which everyone agrees. The difficulties in finding an equilibrium message in this framework starting from a disequilibrium initial position have already been noted (see, for instance, Arrow and Hahn, 1971). First, if no trade is allowed out of equilibrium, then no trade can ever take place if the tatonnement fails to converge (as in Scarf's example, 1960), or, if it does not converge in finite time. Secondly, some related adjustment processes have their difficulties from the point of view of "decentralization" or of mimicking the "invisible hand" (see the assessments of Hahn, 1982, and Smale, 1986). However, if the auctioneer happens to present an equilibrium message (price and allocation) to begin with, the above verification scenario can certainly be envisaged.

In the intertemporal framework, however, a conceptual problem arises when we try to apply Hurwicz's paradigm directly: at any given time not all agents are present — many are yet to be born. But then, as Hurwicz and Weinberger (1990) (H-W henceforth) point out, this means that the auctioneer/planner has to wait until eternity before *all* agents can verify their conditions for equilibrium, so that no allocation decision can actually be carried out in *any* period. It should be emphasized that *this is a difficulty that persists even when the auctioneer chooses an equilibrium message to begin with.* Hence, a process that requires (in the spirit of tatonnement) waiting till all the verification is complete is really ". . . a prescription for economic paralysis rather than a realistic model for economic behaviour." (H-W, p. 317).

To resolve this issue, H-W define an *evolutionary* process. This is a mechanism in which, if agents up to any finite time respond positively to the verification rules for their part of a proposed plan, then the designer actually carries out that part of the plan. This is done *irrespective of whether future agents verify or fail to verify the rules of the mechanism* for their portion of the designer's proposal. (This appears to be somewhat in contrast to the usual non-tatonnement models, in which actual consumption or production of goods does not take place. Typically, trading means moving to a new position in the Edgeworth box, not a change in the size of the box.)

The main result of H-W is negative; they show the impossibility of achieving optimal outcomes for an aggregative growth model using evolutionary processes, when the optimality criterion is the maximization of the discounted sum of one-period utilities.

Our approach in this paper is somewhat different. Instead of focusing on the possibility (or impossibility) of realizing an optimal plan through a decentralized process, we actually *construct* a decentralized mechanism where decisions are carried out period after period in the evolutionary manner of H-W. (The relevant concepts are formally defined in Section 5.) It is shown that this process has interesting normative properties: the allocation sequence generated by it is intertemporally efficient in the sense of Malinvaud (1953), and also maximizes the long run average utility from consumption. (See

Section 2a for precise definitions of these concepts and Section 6 for the stated result.)

Speaking informally, perhaps the reason why our mechanism succeeds despite the H-W impossibility result is this: the optimality criterion we use leads to many (more precisely, an infinite number of) optimal plans. In contrast, the criterion of H-W typically leads to a unique plan which maximizes the discounted sum of utilities (for instance, when the one-period utility function is strictly concave). Intuitively, one expects that it is easier for an evolutionary mechanism, in which plans are continually revised and updated as new information comes in, to achieve one out of an infinite set of outcomes, as opposed to attaining a unique outcome.

A few remarks on the optimality criteria explored in the context of intertemporal economics might be useful to put our result in proper perspective. One can argue on both philosophical grounds (Rawls, 1971) and economic ones (Pigou, 1928; Ramsey, 1928) that when making decisions, the designer should not favor nearby generations to the detriment of generations far in the future. If we accept these arguments, then we must look for some "undiscounted" optimality criterion for making welfare judgements.

An optimality notion that has appeared in the undiscounted case is the maximization of the "long-run average reward." This criterion was first explored in the operations research and statistics literature (see Howard, 1960, or Blackwell, 1962; later references include Veinott, 1966, and Ross, 1968). There are actually several versions of this criterion (for a discussion, see Flynn, 1976). The one we use requires a program to maximize $\liminf_{T\to\infty} [T^{-1} \sum_{t=1}^{T} u(c_t)]$, where $u(\cdot)$ is the one period utility function. Bhattacharya and Majumdar (1989) have shown the existence of stationary optimal policies under such a criterion, for a very general class of semi-Markov models. In economics, a version of this criterion was suggested by Dasgupta (1964), and was subsequently studied by Jeanjean (1974) and Dutta (1986, 1989). Dutta, in particular, has analyzed it in detail, and outlines its relationships with alternative undiscounted optimality notions. We should stress that this criterion has its drawbacks, a major one being that there is no "weight" on the consumption sequence for *any* finite length of time. Consequently, we require an optimal program to maximize the long-run average utility and — in addition — to be efficient.

The mechanism that we construct works via a continual planning revision process (see Goldman, 1968). Roughly speaking this process can be described as follows: given an initial stock $\mathbf{x} > 0$, a 2-period utility maximization problem is considered where the terminal stock is set equal to **x**. The maximal first period consumption (say c_1) takes place and the economy moves to the stock level $x_1 = f(\mathbf{x}) - c_1$, where $f(\cdot)$ is the production function. Next period another 2-period utility maximization problem is contemplated with initial stock and terminal stock set equal to x_1. Again, the maximal consumption takes place and the process is repeated. The sequence of stocks generated in this way is called a "rolling plan." In Section 3, monotonicity and asymptotic properties of such plans are studied. In Section 4, these properties are used to establish that rolling plans are optimal.

In Section 6 we formally verify that a rolling plan can be achieved by a decentralized evolutionary mechanism. Decentralization of decision-making is achieved by introducing some accounting prices. The consumer is required to equate a specific marginal rate of intertemporal substitution with a price ratio and the producer is required to verify feasibility and a condition of intertemporal profit maximization. We note that the above pricing scheme leads to the intertemporal profit maximizing shadow prices that Malinvaud (1953) was concerned with. However, one should observe that it is different from the dual prices in "optimal growth theory," used by Gale (1967) and Gale and Sutherland (1968). The difference is due to the fact that even though the consumer in any period equates his marginal rate of substitution with the accounting price-ratio, only the current part of his resulting 2-period plan is carried out. This difference is perhaps the most significant economic feature of our scheme when contrasted with the earlier duality theory.

2. The Framework

2a. Plans

We consider a one-good model, with a technology given by a production function f from $\Re_+$ to itself.

We define a *plan* from $\mathbf{x} \geq 0$ as a sequence (x_t) satisfying

$$x_0 = \mathbf{x},\ 0 \leq x_t \leq f(x_{t-1}) \quad \text{for } t \geq 1 \ .$$

The consumption sequence (c_t) generated by a plan (x_t) is given by

$$c_t = f(x_{t-1}) - x_t \quad \text{for } t \geq 1 \ .$$

A plan (x_t) from $\mathbf{x} \geq 0$ is called *interior* if $x_t > 0$ and $c_{t+1} > 0$ for $t \geq 0$.

A plan (x_t) from $\mathbf{x} \geq 0$ is *inefficient* if there is a plan (x'_t) from

$\mathbf{x} \geq 0$ satisfying $c'_t \geq c_t$ for all $t \geq 1$ and $c'_t > c_t$ for some t. It is *efficient* if it is not inefficient.

The following assumptions on f are used:

(A.1) $f(0) = 0$.
(A.2) f is continuous on $\Re_+$ and twice continuously differentiable on $\Re_{++}$.
(A.3) f is strictly increasing on $\Re_+$, with $f'(x) > 0$ for $x > 0$.
(A.4) f is strictly concave on $\Re_+$, with $f''(x) < 0$ for $x > 0$.
(A.5) f satisfies the end-point conditions: $f'(x) \to d < 1$ as $x \to \infty$; $f'(x) \to \infty$ as $x \to 0$.

Under (A.1)–(A.5), there exist unique numbers $\hat{k}$, k satisfying $0 < \hat{k} < k < \infty$, $f'(\hat{k}) = 1$, $f(k) = k$. We note that $\hat{k}$ satisfies the inequality $f(\hat{k}) > \hat{k}$. We denote $[f(\hat{k}) - \hat{k}]$ by $\hat{c}$. Thus the sequence (x_t) defined by $x_t = \hat{k}$ for $t \geq 0$ is a plan from $\hat{k}$. We refer to $\hat{k}$ as the *golden-rule input stock* and to $(\hat{k})$ as the *golden-rule plan*. The consumption sequence associated with the golden-rule plan is $(\hat{c})$. We refer to $\hat{c}$ as the *golden-rule consumption*. For any plan (x_t) from $\mathbf{x} \in (0, k)$, it can be shown that $(x_t, c_t) \ll (k, k)$ for $t \geq 1$. We refer to k as the *maximum sustainable input stock*.

Preferences are represented by a utility function u, from $\Re_+$ to $\Re$.

We will say that a plan (x_t) from $\mathbf{x} \geq 0$ *maximizes the long-run average utility* if

$$\liminf_{T\to\infty} \left[\sum_{t=1}^{T} u(c_t)/T\right] \geq \liminf_{T\to\infty} \left[\sum_{t=1}^{T} u(c'_t)/T\right]$$

for every plan (x'_t) from $\mathbf{x}$. A plan (x_t) from $\mathbf{x} \geq 0$ is *optimal* if it maximizes the long-run average utility and, in addition, is efficient.

The following assumptions on u will be used:

(A.6) u is continuous on $\Re_+$ and twice continuously differentiable on $\Re_{++}$.
(A.7) u is strictly increasing on $\Re_+$, with $u'(c) > 0$ for $c > 0$.
(A.8) u is strictly concave on $\Re_+$, with $u''(c) < 0$ for $c > 0$.
(A.9) u satisfies the end-point condition: $u'(c) \to 0$ as $c \to \infty$.

2b. Finite-Horizon Plans

In this sub-section, we will describe what are commonly referred to as "finite-horizon plans." We will focus on those finite-horizon plans for which the terminal (end of horizon) input stock is the same as the initial (beginning of horizon) input stock. This provides the motivation for the formal definition that follows.

Let T be a positive integer greater than one. A *T-plan* from $\mathbf{x} \in (0, k)$ is a finite sequence $(x_t)_{t=0}^T$ satisfying

$$x_T = x_0 = \mathbf{x},\ 0 \leq x_t \leq f(x_{t-1}) \quad \text{for } 1 \leq t \leq T\ .$$

The finite consumption sequence $(c_t)_{t=1}^T$ generated by a T-plan $(x_t)_{t=0}^T$ is given by

$$c_t = f(x_{t-1}) - x_t \quad \text{for } 1 \leq t \leq T\ .$$

If $(x_t)_{t=0}^T$ is a T-plan from $\mathbf{x} \in (0, k)$, it can be checked that $(x_t, c_t) \ll (k, k)$ for $1 \leq t \leq T$.

A *maximal T-plan* from $\mathbf{x} \in (0, k)$ is a T-plan $(x_t^*)_{t=0}^T$ from $\mathbf{x}$ satisfying

$$\sum_{t=1}^{T} u(c_t^*) \geq \sum_{t=1}^{T} u(c_t)$$

for every T-plan $(x_t)_{t=0}^T$ from $\mathbf{x}$.

Given $\mathbf{x} \in (0, k)$, we can define

$$C(\mathbf{x}) = \{(c_t)_{t=1}^T\colon\ (c_t)_{t=1}^T \text{ is a consumption sequence generated by a } T\text{-plan } (x_t)_{t=0}^T \text{ from } \mathbf{x}\}\ .$$

Then it can be checked that $C(\mathbf{x})$ is a non-empty, closed and bounded subset of $\Re_+^T$. We can define $U\colon\ \Re_+^T \to \Re$ by

$$U(c_1, \ldots, c_t) = \sum_{t=1}^{T} u(c_t)\ .$$

Then U is continuous on $\Re_+^T$ and therefore on $C(\mathbf{x})$. Using the Weierstrass theorem, there is $(c_t^*)_{t=1}^T$ in $C(\mathbf{x})$ which maximizes U among all $(c_t)_{t=1}^T$ in $C(\mathbf{x})$. That is, there exists a maximal T-plan $(x_t^*)_{t=0}^T$ from $\mathbf{x}$. Using the concavity of f, $C(\mathbf{x})$ is a convex set, and using the strict concavity of u, U is strictly concave on $C(\mathbf{x})$. Thus U is maximized on $C(\mathbf{x})$ at a unique point, $(c_t^*)_{t=1}^T$. Given the definition of a T-plan, this also means that there is a unique maximal T-plan $(x_t^*)_{t=0}^T$ from $\mathbf{x}$.

If $(x_t^*)_{t=0}^T$ is the maximal T-plan from $\mathbf{x} \in (0, k)$, then using the end-point condition on u, it can be checked that $c_t^* > 0$ for

$t = 1, \ldots, T$. Using $f(0) = 0$, it then follows that $x_t^* > 0$ for $t = 0, \ldots, T$. For $1 \leq t \leq T - 1$, the expression

$$V(x) \equiv u(f(x_{t-1}^*) - x) + u(f(x) - x_{t+1}^*)$$

must be maximized at $x = x_t^*$ among all $x \geq 0$ which satisfy $f(x_{t-1}^*) - x \geq 0$ and $f(x) - x_{t+1}^* \geq 0$. Since $x_t^* > 0$, $c_t^* > 0$ and $c_{t+1}^* > 0$, the maximum is attained at an interior point, so that

$$V'(x_t^*) = u'(c_t^*)(-1) + u'(c_{t+1}^*)f'(x_t^*) = 0 .$$

This yields the well-known *Ramsey-Euler equations*:

$$u'(c_t^*) = u'(c_{t+1}^*)f'(x_t^*) \quad \text{for } 1 \leq t \leq T - 1 .$$

In Sections 4 and 6 of this paper, we will be concerned with T-plans for the special case of $T = 2$; we refer to these naturally as 2-plans. A 2-plan from $\mathbf{x} \in (0, k)$ can be described by the input sequence $(\mathbf{x}, x_1, \mathbf{x})$, with associated consumption sequence $(c_1, c_2) \geq 0$ given by $c_1 = f(\mathbf{x}) - x_1$ and $c_2 = f(x_1) - \mathbf{x}$. We note here, for ready reference, a convenient characterization of maximal 2-plans.

Proposition 2.1: Let $(\mathbf{x}, x_1, \mathbf{x})$ be a 2-plan from $\mathbf{x} \in (0, k)$, with $c_1 \equiv f(\mathbf{x}) - x_1 > 0$ and $c_2 \equiv f(x_1) - \mathbf{x} > 0$. Then $(\mathbf{x}, x_1, \mathbf{x})$ is a maximal 2-plan if and only if

$$u'(f(\mathbf{x}) - x_1) = u'(f(x_1) - \mathbf{x})\, f'(x_1) .$$

Proof: Clearly, necessity follows from our above discussion showing that maximal T-plans satisfy the Ramsey-Euler equations.

For the sufficiency part, let $(\mathbf{x}, x_1, \mathbf{x})$ be any 2-plan from $\mathbf{x}$, with associated consumption sequence (c_1', c_2') defined by $c_1' = f(\mathbf{x}) - x_1'$, $c_2' = f(x_1') - \mathbf{x}$. Then $[u(c_1') + u(c_2')] - [u(c_1) + u(c_2)] \leq u'(c_1)(c_1' - c_1) + u'(c_2)(c_2' - c_2) = u'(c_2)[f'(x_1)(c_1' - c_1) + (c_2' - c_2)] = u'(c_2)[f'(x_1)(x_1 - x_1') + (f(x_1') - f(x_1))] \leq u'(c_2)[f'(x_1)(x_1 - x_1') + f'(x_1)(x_1' - x_1)] = 0$. This shows that $(\mathbf{x}, x_1, \mathbf{x})$ is a maximal 2-plan from $\mathbf{x}$.

2c. Rolling Plans

Rolling plans are defined in terms of "finite-horizon plans." We will focus on rolling plans "generated by" those finite-horizon plans for which the terminal input stock is the same as the initial input stock; that is by those finite horizon plans which we referred to as T-plans in the previous sub-section.

Using the results of the previous sub-section, we can conclude that there is a function h from $(0, k)$ to $\Re_+$, such that if $x \in (0, k)$, and $(x_t)_{t=1}^T$ is the unique maximal T-plan from x, then $x_1 = h(x)$. We might further observe that given any $x \in (0, k)$, $0 < h(x) < f(x) < k$, so that h is a function from $(0, k)$ to $(0, k)$.

A *rolling plan* from $\mathbf{x} \in (0, k)$ is a sequence (x_t) satisfying

$$x_0 = x, \; x_{t+1} = h(x_t) \quad \text{for } t \geq 0 \; .$$

Notice that the sequence (x_t) is well defined, since h maps $(0, k)$ to $(0, k)$. Furthermore (x_t) is a plan from $\mathbf{x}$, since $h(x_t) \leq f(x_t)$ for $t \geq 0$. Thus it has an associated consumption sequence (c_t) defined by

$$c_t = f(x_{t-1}) - h(x_{t-1}) \quad \text{for } t \geq 1 \; .$$

We refer to h as the "generating function" of rolling plans.

In view of Proposition 2.1, if $T = 2$, and h is the generating function of rolling plans then for every $x \in (0, k)$, $u'(f(x) - h(x)) = u'(f(h(x)) - x)f'(h(x))$.

3. Monotone Convergence Properties of Rolling Plans

Rolling plans can be shown to be monotone in input stocks over time. If a rolling plan starts from an initial input stock below the golden-rule input stock ($\hat{k}$), then the input stocks monotonically increase and converge to the golden-rule input stock. (An analogous statement can be made if the initial input stock is above the golden-rule input stock.) Such properties were established by Goldman (1968) in the context of a continuous-time aggregative model (with discounting of future utilities). In discrete-time aggregative models, similar properties can be established by focusing on the properties of what we have called the "generating function" of rolling plans. We follow this method in this section. Specifically, Lemma 3.1 shows that the generating function is above (below) the 45-degree line at input stocks below (above) the golden-rule level; Lemma 3.2 shows that the generating function is also increasing on its domain. These properties are then summarized in Proposition 3.1. Proposition 3.2 establishes that these properties imply monotone convergence of rolling plans to the golden-rule.

Lemma 3.1: If $(x_t^*)_{t=0}^T$ is the maximal T-plan from $\mathbf{x} \in (0, k)$, then (a) $\mathbf{x} < \hat{k}$ implies $\mathbf{x} < x_1^* < \hat{k}$; (b) $\mathbf{x} = \hat{k}$ implies $x_1^* = \hat{k}$; (c) $\mathbf{x} > \hat{k}$ implies $\mathbf{x} > x_1^* > \hat{x}$.

Proof: We will prove (a); the proofs of (b) and (c) can be worked out analogously. We first establish that $x_1^* < \hat{k}$. Suppose instead that $x_1^* \geq \hat{k}$. Then $f'(x_1^*) \leq 1$, so by the Ramsey-Euler equation, $u'(c_1^*) = f'(x_1^*)u'(c_2^*) \leq u'(c_2^*)$. Since $u''(c) < 0$ for $c > 0$, we get $c_1^* \geq c_2^*$. Thus

$$f(\mathbf{x}) - x_1^* \geq f(x_1^*) - x_2^* > f(\mathbf{x}) - x_2^* ,$$

the last inequality following from $x_1^* \geq \hat{k} > \mathbf{x}$, and the fact that f is increasing. Thus, $x_2^* > x_1^* \geq \hat{k}$. We can then repeat the steps to obtain

$$x_T^* > x_{T-1}^* > \ldots > x_2^* > x_1^* > \mathbf{x} ,$$

so that $x_T^* > \mathbf{x}$, which contradicts the fact that $x_T^* = \mathbf{x}$ by definition of a T-plan. Thus $x_1^* < \hat{k}$.

Next, we establish that $x_1^* > \mathbf{x}$. Suppose instead that $x_1^* \leq \mathbf{x}$. Since $x_1^* < \hat{k}$, we have $f'(x_1^*) > 1$, so by the Ramsey-Euler equation, $u'(c_1^*) = f'(x_1^*)u'(c_2^*) > u'(c_2^*)$. Thus, $c_1^* < c_2^*$, and

$$f(\mathbf{x}) - x_1^* < f(x_1^*) - x_2^* \leq f(\mathbf{x}) - x_2^* .$$

Hence, $x_2^* < x_1^*$. We can then repeat the steps to obtain

$$x_T^* < x_{T-1}^* < \ldots < x_2^* < x_1^* \leq \mathbf{x} ,$$

so that $x_T^* < \mathbf{x}$, which contradicts the fact that $x_T^* = \mathbf{x}$. Thus $x_1^* > \mathbf{x}$.

Q.E.D.

Lemma 3.2: If $(x_t^*)_{t=0}^T$ is the maximal T-plan from $x^* \in (0, k)$, and $(\bar{x}_t)_{t=0}^T$ is the maximal T-plan from $\bar{x} \in (0, k)$, and $x^* > \bar{x}$, then $x_1^* > \bar{x}_1$.

Proof: Suppose the hypotheses of Lemma 3.2 are valid, but $x_1^* \leq \bar{x}_1$. Then, we have $c_1^* = f(x^*) - x_1^* > f(\bar{x}) - \bar{x}_1 = \bar{c}_1$. Thus, we must have $f'(x_1^*) \geq f'(\bar{x}_1)$ and $u'(c_1^*) < u'(\bar{c}_1)$. Using the Ramsey-Euler equations for the two maximal T-plans, we obtain

$$1 > \frac{u'(c_1^*)}{u'(\bar{c}_1)} = \frac{f'(x_1^*)u'(c_2^*)}{f'(\bar{x}_1)u'(\bar{c}_2)} \geq \frac{u'(c_2^*)}{u'(\bar{c}_2)} .$$

This means $u'(c_2^*) < u'(\bar{c}_2)$, and so $c_2^* > \bar{c}_2$. Thus, we obtain

$$f(x_1^*) - x_2^* > f(\bar{x}_1) - \bar{x}_2 \geq f(x_1^*) - \bar{x}_2 ,$$

so that $\bar{x}_2 > x_2^*$. The above argument can then be repeated to get $\bar{x}_t > x_t^*$ for $t = 2, \dots, T$. Thus, by definition of T-plans, we obtain $x^* = x_T^* < \bar{x}_T = \bar{x}$, which contradicts the hypothesis that $x^* > \bar{x}$.

Q.E.D.

Proposition 3.1: The generating function, h, has the following properties: (a) for $x \in (0, k)$, $0 < h(x) < f(x) < k$; (b) for $x \in (0, k)$, $\hat{k} \gtreqless h(x) \gtreqless x$ as $\hat{k} \gtreqless x$; (c) h is increasing on $(0, k)$; (d) h is continuous on $(0, k)$; (e) $\lim_{x \to 0} h(x) = 0$; (f) $\lim_{x \to k} h(x) = k$.

Proof: Clearly, (a) follows from our discussion in Section 2c. Also (b) follows from Lemma 3.1 and (c) follows from Lemma 3.2.

To establish (d), we proceed to apply the Maximum Theorem (Berge, 1963, p. 116). Define $D = (0, k)$, $\bar{D} = [0, k]$; then $\bar{D}^T$ is a compact subset of $\Re^T$, and D is a subset of $\Re$. Note that $U(c_1, \dots, c_T)$ is a continuous function from $\bar{D}^T$ to $\Re$ (by continuity of u). Also $C(x)$ is a continuous correspondence from D to $\bar{D}^T$ (for definitions of $U(c_1, \dots, c_T)$ and $C(x)$, see Section 2b). To see this last assertion, note that $C(x)$ is clearly an upper semicontinuous correspondence from D to $\bar{D}^T$, by continuity of f. To check lower semicontinuity of $C(x)$, let $x^s \in D$ for $s = 1, 2, \dots$, with $x^s \to \bar{x} \in D$, and let $(\bar{c}_1, \dots, \bar{c}_T) \in C(\bar{x})$. Then there is a T-plan $(\bar{x}_t)_{t=0}^T$ from $\bar{x}$, such that $(\bar{c}_t)_{t=1}^T$ is the consumption sequence generated by it. From our discussion of Section 2b, we know that $(\bar{x}_t, \bar{c}_t) \gg 0$ for $t = 1, \dots, T$. Thus (using the continuity of f) we can pick $\epsilon > 0$ such that $1 - \epsilon > 0$, $\bar{x}(1 + \epsilon) < k$, and for all $\lambda \in [1 - \epsilon,\ 1 + \epsilon]$, $f(\lambda \bar{x}_t) - \lambda \bar{x}_{t+1} > 0$ for $t = 0, \dots, T - 1$. Since $x^s \to x$ as $s \to \infty$, there is some $\bar{s}$, such that for $s \geq \bar{s}$, $(x^s / \bar{x}) \in [1 - \epsilon,\ 1 + \epsilon]$. For $s \geq \bar{s}$, define $\lambda^s = (x^s / \bar{x})$, and a sequence $(x_t^s)_{t=0}^T$ by $x_t^s = \lambda^s \bar{x}_t$ for $t = 0, \dots, T$. Then by construction, $(x_t^s)_{t=0}^T$ is a T-plan for each $s \geq \bar{s}$, and its associated consumption sequence $(c_t^s)_{t=1}^T$ satisfies $c_t^s = f(\lambda^s \bar{x}_t) - \lambda^s \bar{x}_{t+1}$ for $t = 1, \dots, T$. As $s \to \infty$, we have $x^s \to \bar{x}$, $\lambda^s \to 1$ and $c_t^s \to \bar{c}_t$ for $t = 1, \dots, T$, by continuity of f. This establishes lower semicontinuity of $C(x)$ on D. Applying the Maximum Theorem, and denoting by $(c_1(x), \dots, c_T(x))$ the (unique) maximizer of U on $C(x)$ for each $x \in D$, we note that $(c_1(x), \dots, c_T(x))$ is a continuous function from D to $\bar{D}_T$. Denoting by $(x_t(x))_{t=0}^T$ the T-plan from x, with associated consumption sequence $(c_t(x))_{t=1}^T$, we note that $(x_1(x), \dots, x_T(x))$ is also a continuous function from D to $\bar{D}^T$. In particular then, $h(x) \equiv x_1(x)$ is a continuous function on D.

We can establish (e) as follows. For $0 < x < \hat{k}$, we have, by (a) and (b), $x < h(x) < f(x)$. Thus as $x \to 0$, $f(x) \to 0$ by continuity of f and $f(0) = 0$, so that $h(x) \to 0$.

If (f) were violated, then there would exist a sequence (x^s), $s = 1, 2, \ldots$, such that $x^s \to k$ as $s \to \infty$, and $h(x^s) \to k' < k$ (using (c), $k' \geq \hat{k}$). Clearly, $f^{T-1}(k') < k$ (where f^{T-1} is the $(T-1)$ iteration of the function, f), and so we must have, for s large, $f^{T-1}(h(x^s)) < x^s$, which contradicts the definition of h.

Q.E.D.

Proposition 3.2: If (x_t) is a rolling plan from $\mathbf{x} \in (0, k)$, then (a) $\mathbf{x} < \hat{k}$ implies that x_t monotonically increases to $\hat{k}$ as $t \to \infty$; (b) $\mathbf{x} > \hat{k}$ implies that x_t monotonically decreases to $\hat{k}$ as $t \to \infty$; (c) $\mathbf{x} = \hat{k}$ implies that $x_t = \hat{k}$ for all $t \geq 1$.

Proof: We will establish (a); (b) and (c) can be proved in a similar manner. If $\mathbf{x} < \hat{k}$, then $\mathbf{x} < h(\mathbf{x}) < \hat{k}$ by Proposition 3.1 (b), so that $\mathbf{x} < x_1 < \hat{k}$. Using Proposition 3.1 (b) again, $x_1 < h(x_1) < \hat{k}$, so that $x_1 < x_2 < \hat{k}$. Repeating this step, we see that x_t monotonically increases while remaining bounded above by $\hat{k}$. Hence, it converges to some k^*, satisfying $0 < k^* \leq \hat{k}$. Using the fact that $x_t \leq h(x_t) \leq x_{t+2}$ for $t \geq 2$, and Proposition 3.1 (d), $h(k^*) = k^*$, so that by Proposition 3.1 (b), $k^* = \hat{k}$.

Q.E.D.

We conclude this section by presenting an example, where the generating function can be numerically computed.

Example 3.1: Let the production function be given by

$$f(x) = 2x^{1/2} \quad \text{for } x \geq 0 .$$

Then f satisfies (A.1)–(A.5); the golden-rule input stock $\hat{k} = 1$, and the maximum sustainable input stock $k = 4$. Let the utility function be given by

$$u(c) = c^{1/2} \quad \text{for } c \geq 0 .$$

Then u satisfies (A.6)–(A.9). Let the planning horizon be fixed at $T = 2$.

A 2-plan from $x \in (0, 4)$ is then a vector (x, x_1, x), with $0 \leq x_1 \leq f(x)$ and $x \leq f(x_1)$. If (x, x_1, x) is a maximal 2-plan from $x \in (0, 4)$, then using the Ramsey-Euler equations, we get

$$2x^{1/2} - x_1 + xx_1 = 2x_1^{3/2} .$$

Denoting $x^{1/2}$ by β, and $x_1^{1/2}$ by α, we get

$$2\alpha^3 + (1 - \beta^2)\,\alpha^2 - 2\beta = 0 \ , \tag{1}$$

where $\beta \in (0, 2)$.

Given $\beta \in (0, 2)$, equation (1) is a cubic in α. Since it is of odd degree, with the last coefficient negative ($-2\beta < 0$) and first coefficient positive ($2 > 0$) it has *at least* one positive real root. On the other hand, by Descartes' rule of signs, it has *at most* one positive real root (since regardless of the sign of $(1 - \beta^2)$, there is exactly one change of sign in the equation). Thus, there is exactly one positive real root to equation (1). If we call this root $\phi(\beta)$, then the generating function, h, is given by

$$h(x) = [\phi(x^{1/2})]^2 \ . \tag{2}$$

While our interest is naturally in the unique positive root of equation (1), we note that *all* the roots of the equation can be found by the standard Cardan-Tartaglia method or the trigonometric method, depending on the sign of the discriminant (see Birkhoff and MacLane, 1977, chs. 4, 5 for details).

The graph of the generating function defined in (2) can be numerically computed and is shown in Figure 1.

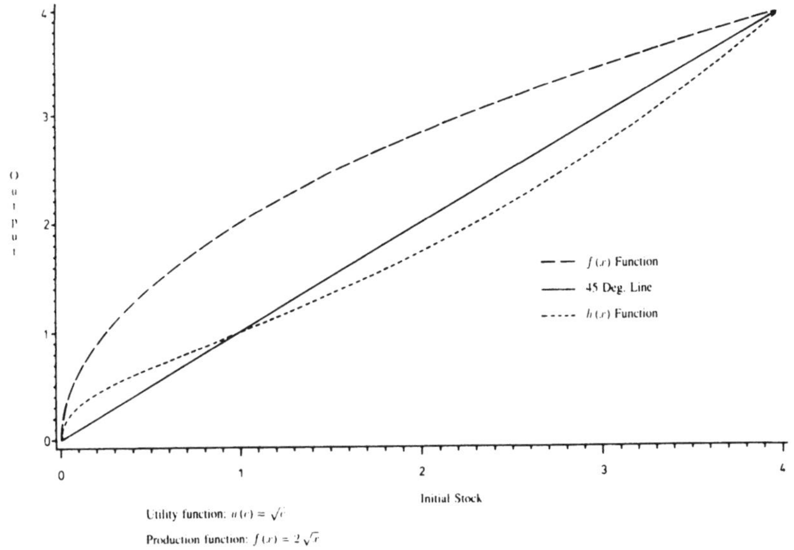

Fig. 1. Rolling plan generating function

4. Welfare Properties of Rolling Plans

The purpose of this section is to show that rolling plans are optimal according to the criterion given in Section 2a. To this end, we establish that rolling plans are efficient (Theorem 4.1), and "good" in the sense of Gale (1967) (Theorem 4.2). Both results are crucially dependent on the fact that the sequence of input stocks of a rolling plan converges to the golden-rule input stock at a geometric rate (Lemma 4.2). This result, in turn, is derived from the property of the generating function, that it has a derivative at the golden-rule input stock with a value strictly between zero and one (Lemma 4.1). Since good plans always maximize the long-run average utility, the above results can be combined to establish the optimality of rolling plans (Theorem 4.3).

In what follows, we fix the time horizon of T-plans at $T = 2$. In the more general case ($T \geq 2$), the result can be obtained by using monotone properties of maximal T-plans with respect to the initial stock and the length of the horizon.

We first introduce some notation which will ease the writing of our results and proofs. Recall that $\hat{c} \equiv f(\hat{k}) - \hat{k} > 0$. We write

$$\begin{aligned} \theta &= \min\left[\{\hat{c}/4f'(\hat{k}/2)\},\ (\hat{k}/2)\right] , \\ \Theta &= (\hat{k} - \theta,\ \hat{k} + \theta) . \end{aligned} \tag{3}$$

It follows from (3) that $\hat{k} - \theta > 0$, and $0 < \hat{c} - \theta = f(\hat{k}) - (\hat{k} + \theta) < k - (\hat{k} + \theta)$; thus, $\hat{k} + \theta < k$. Consequently the set Θ is an open sub-interval of $(0, k)$. Note that if $(x, z) \in \Theta^2$, then $f(x) - z \geq f(\hat{k} - \theta) - f(\hat{k}) + f(\hat{k}) - (\hat{k} + \theta) \geq f'(\hat{k} - \theta)(-\theta) + \hat{c} - \theta > (3\hat{c}/4) - \theta f'(\hat{k}/2) \geq (\hat{c}/2) > 0$.

Lemma 4.1: There is $\delta \in (0, \theta)$ (where θ is given by (3)) such that the generating function h is continuously differentiable on the set

$$\Delta \equiv [\hat{k} - \delta,\ \hat{k} + \delta] \tag{4}$$

and $0 < h'(x) < 1$ for all $x \in \Delta$.

Proof: Consider the function γ: $\Theta^2 \to \Re$ defined by

$$\gamma(x, z) \equiv u'(f(x) - z) + u'(f(z) - x)f'(z) .$$

Then γ is continuously differentiable on Θ^2. Furthermore, $\gamma(\hat{k}, \hat{k}) = 0$ and $\gamma_z(\hat{k}, \hat{k}) < 0$. So, by the implicit function theorem, there are open neighborhoods N_0 and N_1 of $\hat{k}$ (where both N_0 and N_1 are subsets of Θ), and a unique function L: $N_0 \to N_1$, such that $\hat{k} = L(\hat{k})$, and $\gamma(x, L(x)) = 0$ for all $x \in N_0$. Furthermore L is continuously differentiable on N_0.

Since N_0 and N_1 are open, we can find $\theta > \delta' > 0$ such that $N' \equiv (\hat{k} - \delta', \hat{k} + \delta')$ is a subset of N_0 and N_1. Since $h(\hat{k}) = \hat{k}$, and h is continuous on $(0, k)$ by Proposition 3.1, we can find $0 < \hat{\delta} < \delta'$, such that for $x \in \hat{N} \equiv (\hat{k} - \hat{\delta},\ \hat{k} + \hat{\delta})$, $h(x) \in N'$.

Now, we observe that by the definition of h, $\gamma(x,\, h(x)) = 0$ for all $x \in \hat{N}$ (see Section 2c). By the implicit function theorem (above), this is possible iff $h(x) = L(x)$ for all $x \in \hat{N}$. This proves that in the neighborhood $\hat{N}$ of $\hat{k}$, h is continuously differentiable.

Evaluating the derivative of h on $\hat{N}$,

$$h'(x) = \frac{u''(f(x) - z)f'(x) + u''(f(z) - x)f'(z)}{u''(f(x) - z) + u''(f(z) - x)f'(z)^2 + u'(f(z) - x)f''(z)}\,,$$

where $z = h(x)$. Thus, evaluating the derivative of h at $\hat{k}$, we get

$$h'(\hat{k}) = 2u''(\hat{c})/[2u''(\hat{c}) + u'(\hat{c})f''(\hat{k})]\ .$$

Using the facts that $u'(\hat{c}) > 0$, $f''(\hat{k}) < 0$ and $u''(\hat{c}) < 0$, we obtain $0 < h'(\hat{k}) < 1$.

Since h' is continuous on $\hat{N}$, and $0 < h'(\hat{k}) < 1$, we can find $0 < \delta < \hat{\delta}$ such that on $\Delta \equiv [\hat{k} - \delta,\ \hat{k} + \delta]$, we have $0 < h'(x) < 1$. Clearly, $\delta \in (0, \theta)$.

Q.E.D.

Lemma 4.2: Suppose (x_t) is a rolling plan from $x \in (0, k)$. Then there is $A > 0$, a positive integer S, and $\rho \in (0, 1)$, such that $x_t \in \Delta$ for $t \geq S$ (where Δ is given by Lemma 4.1), and

$$|\, x_t - \hat{k}\,| \leq A\rho^t \quad \text{for } t \geq S\ .$$

Proof: Consider the interval Δ obtained in Lemma 4.1. We know that h' is continuous on Δ, and $0 < h'(x) < 1$ for all $x \in \Delta$. Let ρ be the maximum value of h' on Δ; then $0 < \rho < 1$.

By Proposition 3.2, there is a positive integer S, such that $x_t \in \Delta$ for all $t \geq S$. Since h is continuously differentiable on Δ, we can use the Mean Value theorem for each $t \geq S$ to obtain

$$|\, x_{t+1} - \hat{k}\,| = |\, h(x_t) - h(\hat{k})\,| = |\, h'(z_t)\,|\,|\, x_t - \hat{k}\,|\ ,$$

where z_t is between x_t and $\hat{k}$. Since $z_t \in \Delta$, we obtain $0 < h'(z_t) \leq \rho < 1$. Thus, for each $t \geq S$,

$$|x_{t+1} - \hat{k}| \leq \rho |x_t - \hat{k}| .$$

Iterating this inequality we obtain for $t \geq S$

$$|x_t - \hat{k}| \leq \rho^{t-S} |x_s - \hat{k}| \leq \rho^{t-S} \delta ,$$

where δ is given by Lemma 4.1. Defining $A = (\delta/\rho^S)$, we have $|x_t - \hat{k}| \leq A\rho^t$ for $t \geq S$, which proves the Lemma.

Q.E.D.

Theorem 4.1: Suppose (x_t) is a rolling plan from $x \in (0, k)$. Then (x_t) is efficient.

Proof: If $x \leq \hat{k}$, then by Proposition 3.2, $x \leq x_t \leq \hat{k}$ for all $t \geq 0$. Thus, using the characterization of efficiency given in Cass (1972), (x_t) is efficient.

If $x > \hat{k}$, we proceed as follows. Consider the set $\Delta = [\hat{k} - \delta,\ \hat{k} + \delta]$ obtained in Lemma 4.1. Using Lemma 4.2, we can find $A > 0$, a positive integer S, and $\rho \in (0, 1)$, such that $x_t \in \Delta$ for $t \geq S$, and

$$|x_t - \hat{k}| \leq A\rho^t \quad \text{for } t \geq S .$$

Let m be the maximum value of $[-f''(x)]$ on Δ. Choose $s \geq S$, such that $[mA\rho^s/(1-\rho)] \leq (1/2)$. Now, for $t \geq s$, we have $x_t \in \Delta$, and so by the Mean Value theorem, $f'(x_t) - f'(\hat{k}) = f''(z_t)(x_t - \hat{k})$, where $x_t \geq z_t \geq \hat{k}$. Since $z_t \in \Delta$, we have $[-f''(z_t)] \leq m$. Thus, for $t \geq s$, $f'(x_t) \geq 1 - m(x_t - \hat{k}) \geq 1 - mA\rho^t = 1 - (mA\rho^s)\rho^{t-s} \geq 1 - [(1-\rho)\rho^{t-s}/2]$. Using this information, we obtain for $t \geq s$,

$$\begin{aligned}
\pi_{t+1} = \prod_{n=0}^{t} f'(x_n) &\geq \pi_s \prod_{n=s}^{t} \{1 - [(1-\rho)\rho^{n-s}/2]\} \\
&\geq \pi_s \{1 - \sum_{n=s}^{t} [(1-\rho)\rho^{n-s}/2]\} \\
&\geq \pi_s \{1 - \sum_{n=0}^{\infty} [(1-\rho)\rho^{n}/2]\} \\
&= \pi_s \{1 - (1/2)\} = (1/2)\pi_s .
\end{aligned}$$

Thus, $\sum_{n=0}^{\infty} \pi_{t+1}$ is divergent. Since $x_t \geq \hat{k}$ for $t \geq 0$, we can again use the characterization of efficiency in Cass (1972) to conclude that (x_t) is efficient.

Q.E.D.

Following Gale (1967), we define a plan (x_t) to be *good* if there is a real number B such that

$$\sum_{t=1}^{\tau}[u(c_t) - u(\hat{c})] \geq B \quad \text{for all } \tau \geq 1 \; .$$

Theorem 4.2: Suppose (x_t) is a rolling plan from $x \in (0, k)$. Then (x_t) is good.

Proof: Using Lemma 4.2, there is $A > 0$, a positive integer S and $\rho \in (0, 1)$, such that $x_t \in \Delta$ (where Δ is given by Lemma 4.1) and $|\, x_t - \hat{k}\,| \leq A\rho^t$ for $t \geq S$. Let b be the maximum value of f' on Δ. For $t \geq S$, $(x_t,\; x_{t+1})$ is in Δ^2 and so in Θ^2, where Θ is given by (3). Thus $f(x_t) - x_{t+1} \geq (\hat{c}/2)$ for $t \geq S$.

Consider first the case where $x < \hat{k}$. In this case, by Proposition 3.2, $0 \leq (\hat{x} - x_t) \leq A\rho^t$ for $t \geq S$. Then, for $t \geq S$, we obtain $\hat{c} - c_{t+1} = [f(\hat{k}) - \hat{k}] - [f(x_t) - x_{t+1}] \leq [f(\hat{k}) - f(x_t)] \leq f'(x_t)[\hat{k} - x_t] \leq bA\rho^t$. This information yields for $t \geq S$, $u(\hat{c}) - u(c_{t+1}) \leq u'(c_{t+1})(\hat{c} - c_{t+1}) \leq u'(\hat{c}/2)bA\rho^t$. Summing this inequality from S to ∞, we get

$$\sum_{t=S}^{\infty}[u(\hat{c}) - u(c_{t+1})] \leq u'(\hat{c}/2)bA\rho^S/(1 - \rho) \; .$$

It follows immediately that (x_t) is good.

Consider next the case where $x \geq \hat{k}$. By Proposition 3.2, $0 \leq (x_t - \hat{x}) \leq A\rho^t$ for $t \geq S$. Then for $t \geq S$, we obtain $\hat{c} - c_{t+1} = [f(\hat{k}) - \hat{k}] - [f(x_t) - x_{t+1}] \leq (x_{t+1} - \hat{k}) \leq A\rho^{t+1}$. This yields for $t \geq S$

$$u(\hat{c}) - u(c_{t+1}) \leq u'(c_{t+1})(\hat{c} - c_{t+1}) \leq u'(\hat{c}/2)A\rho^{t+1} \; .$$

Then, following the above procedure, (x_t) is good.

Q.E.D.

Theorem 4.3: Suppose (x_t) is a rolling plan from $x \in (0, k)$. Then (x_t) is optimal.

Proof: Recall that $\hat{c} = f(\hat{k}) - \hat{k}$ is the golden-rule consumption. Denote $u'(\hat{c})$ by $\hat{p}$. Let (x'_t) be any plan from x. Then, for $t \geq 1$, $[u(c'_t) - u(\hat{c})] \leq \hat{p}[c'_t - \hat{c}] = \hat{p}[\{f(x'_{t-1}) - x'_t\} - \{f(\hat{k}) - \hat{k}\}] =$

$\hat{p}[f(x'_{t-1}) - f(\hat{k})] - \hat{p}[x'_t - \hat{k}] \leq \hat{p}[x'_{t-1} - \hat{k}] - \hat{p}[x'_t - \hat{k}]$, using the concavity of f and $f'(\hat{k}) = 1$. Thus, summing this inequality from $t = 1$ to $t = \tau$ we get

$$\sum_{t=1}^{\tau}[u(c'_t) - u(\hat{c})] \leq \hat{p}x \ .$$

This yields for all $\tau \geq 1$

$$(1/\tau)\sum_{t=1}^{\tau} u(c'_t) \leq u(\hat{c}) + (\hat{p}x/\tau) \ .$$

Thus, taking the inferior limit of both sides,

$$\liminf_{\tau\to\infty} (1/\tau)\sum_{t=1}^{\tau} u(c'_t) \leq u(\hat{c}) \ . \tag{5}$$

Using Theorem 4.2 above, we know that (x_t) is good. Thus, there is some real number B such that for all $\tau \geq 1$,

$$\sum_{t=1}^{\tau}[u(c_t) - u(\hat{c})] \geq B \ .$$

This yields for all $\tau \geq 1$

$$(1/\tau)\sum_{t=1}^{\tau} u(c_t) \geq u(\hat{c}) + (B/\tau) \ .$$

Again, taking the inferior limit of both sides,

$$\liminf_{\tau\to\infty} (1/\tau)\sum_{t=1}^{\tau} u(c_t) \geq u(\hat{c}) \ . \tag{6}$$

Inequalities (5) and (6) clearly imply that the rolling plan (x_t) maximizes long-run average utility. Using Theorem 4.1 above, (x_t) is also efficient. Hence, (x_t) is optimal.

Q.E.D.

Remarks:

(i) The fact that a good plan maximizes long-run average utility among all plans has already been noted in the literature (see, for example, Jeanjean, 1974). We have given the proof here for the sake of completeness.

(ii) A plan which maximizes long-run average utility need not be good. Consider $0 < x < \hat{k}$ satisfying $f(x) > \hat{k}$, and a sequence (x_t) defined by $x_t = x$ for $t = 2^n$ $(n = 0, 1, 2, \ldots)$ and $x_t = \hat{k}$ for $t \neq 2^n$. It can be checked that (x_t) is a plan from x which maximizes long-run average utility and is also efficient but is not good. Thus, Theorems 4.1 and 4.2 actually establish a stronger welfare result about rolling plans than is reflected in Theorem 4.3.

(iii) The properties of efficiency and "goodness" of a plan are independent of each other. An efficient plan need not be good (see the example in remark (ii) above). Similarly, a good plan need not be efficient. Consider $\hat{k} < x < k$, and a sequence (x_t) defined by $x_0 = x$, $x_{t+1} = f(x_t) - \hat{c}$ for $t \geq 0$. It can be checked that (x_t) is a plan. It is clearly good, since $c_t = \hat{c}$ for all $t \geq 1$. It is also clearly inefficient since the sequence (x'_t) defined by $x'_0 = x$, $x'_t = \hat{k}$ for $t \geq 0$ is a plan from x with $c_1 > \hat{c}$ and $c_t = \hat{c}$ for all $t \geq 2$.

5. Decentralized Evolutionary Mechanisms

5a. Evolutionary Mechanisms

In the rest of the paper, we are primarily concerned with the problem of realizing optimal allocations in our intertemporal economy with the help of a suitably constructed "decentralized evolutionary mechanism." The notion of a mechanism is by now a familiar one (see, for instance, Mount and Reiter, 1974, and Hurwicz, 1986, for discussions), but we will have to be careful in defining an intertemporal version of it, if we want to capture the special structure of sequential decision-making that is involved. This section is entirely devoted to this task.

We first define formally what we mean by an "evolutionary mechanism." This is followed by an informal discussion of how this mechanism is supposed to operate.

We consider the following objects to be given: a *set of environments* $E \subset \prod_{t=1}^{\infty} E_t$, a *space of allocations* $A = \prod_{t=1}^{\infty} A_t$, and a *state space* $S = \prod_{t=0}^{\infty} S_t$. We consider A_t to be a subset of a finite dimensional real space $\Re^{\ell}$, and S_t to be a subset of a finite dimensional real space $\Re^q$, for all $t \geq 1$.

An *evolutionary mechanism* is a sequence (M_t, G_t, H_t) where:

(a) M_t, the *message space* in period t, is a subset of a finite dimensional real space, denoted by $\Re^m$;

(b) G_t, the *verification function* in period t, is a mapping from $E_t \times S_{t-1} \times M_t$ to a finite dimensional real space, denoted

by $\Re^n$. It is required to satisfy the following condition for each $(e_t, s_{t-1}) \in E_t \times S_{t-1}$:
There is a unique message $m_t \in M_t$ such that $G_t(e_t, s_{t-1}, m_t) = 0$;

(c) $H_t \equiv (H_t^1, H_t^2)$, the *outcome function* in period t, is a mapping from M_t to $\Re^q \times \Re^\ell$. It is required to satisfy the following condition for each $(e_t, s_{t-1}) \in E_t \times S_{t-1}$:
If $m_t \in M_t$ and $G_t(e_t, s_{t-1}, m_t) = 0$ then $H_t(m_t) \in S_t \times A_t$.

Given an evolutionary mechanism (M_t, G_t, H_t), we define the *equilibrium message function* in period t as

$$\mu_t(e_t, s_{t-1}) = \{m_t \in M_t\colon\ G_t(e_t, s_{t-1}, m_t) = 0\}$$

for each $(e_t, s_{t-1}) \in E_t \times S_{t-1}$. The *equilibrium outcome function* in period t is defined as

$$\nu_t(e_t, s_{t-1}) = \{H_t(m_t)\colon\ m_t \in \mu_t(e_t, s_{t-1})\}$$

for each $(e_t, s_{t-1}) \in E_t \times S_{t-1}$. We refer to ν_t^1 $[\equiv H_t^1(m_t)]$ as the *equilibrium state* in period t, and to ν_t^2 $[\equiv H_t^2(m_t)]$ as the *equilibrium allocation* in period t (where m_t is the equilibrium message in period t).

Several observations about the above definitions are worth making at this point:

(i) First, our definition of an evolutionary mechanism is in the same spirit as the notion of an "evolutionary process" introduced by Hurwicz and Weinberger (1990, p. 317) (see particularly the discussion of this concept on pp. 317–318 of their paper). Both the definitions require that the outcome H_t in period t be independent of verifications in the future (that is, verifications for $t+1$ onwards). However, the exact equivalence of the two notions is not a subject we wish to pursue here.

(ii) From the point of view of applications, we have found it important to bring in explicitly the notion of a state space, and to include the "state variable" as an argument in the verification functions, G_t.

(iii) The dimensionality constraints on the message space $[M_t \subset \Re^m]$ and on the range of the verification functions $[G_t(e_t, s_{t-1}, m_t) \in \Re^n$ for each $(e_t, s_{t-1}, m_t) \in E_t \times S_{t-1} \times M_t]$ reflect the notion that transmission and usage of information is costly and hence that agents can communicate or process only a finite amount of information in each period.

(iv) The condition which G_t is required to satisfy ensures that μ_t and ν_t are well-defined functions. The condition that there is a *unique* equilibrium message for each $(e_t, s_{t-1}) \in E_t \times S_{t-1}$ is surely restrictive. The more general case of an "equilibrium message correspondence" can, of course, be treated, but at the expense of considerable complication which would add little to the applications we have in mind. Thus, we have deliberately kept the strong restriction on the verification function. Since our main interest is in demonstrating a "possibility result," the more demanding are our requirements for an evolutionary mechanism, the stronger is our possibility result.

To see how this mechanism operates, consider that an environment for period 1, e_1, is given. This would typically describe the preferences of consumers and technological possibilities of producers in period 1.

Consider, also, that an initial state, s_0, is given. The initial state would typically be described by the various capital and resource stocks available at the end of period zero.

In period 1, the mechanism designer proposes a message m_1 to agents. The agents (knowing e_1 and s_0) would then verify whether $G_1(e_1, s_0, m_1) = 0$. Notice that agents in period 1 are being required to be informed about the environment in period 1, e_1, as well as the previous period's state, s_0; these appear to be plausible requirements. If $G_1(e_1, s_0, m_1) = 0$, then m_1 is the equilibrium message. (If $G_1(e_1, s_0, m_1) \neq 0$, another message has to be proposed and the process has to be repeated until the equilibrium message is found. How the equilibrium message is found is itself a topic of considerable interest; but we will not be concerned with it here.)

If m_1 is the equilibrium message in period 1, the outcome function H_1^1 specifies a state s_1 in S_1, consisting typically of capital and resource stocks available at the end of period 1 and the outcome function H_1^2 specifies an allocation a_1 in A_1, consisting typically of consumption and investment decisions in period 1. This state, allocation pair is the "equilibrium outcome" of period 1. It is to be understood that the allocation corresponding to this equilibrium outcome is actually carried out; similarly the state corresponding to this equilibrium outcome is actually attained.

In period 2, knowing the state that was actually attained in period 1 (s_1) and the environment in period 2 (e_2), the same procedure yields the equilibrium outcome of period 2, and hence the state at the end of period 2. This step is repeated indefinitely.

The above description of the operation of a mechanism (M_t, G_t, H_t) implies that if an environment $e \in E$ is given, and an initial state $s \in S_0$ is specified, the mechanism then defines uniquely the state and allocation sequence for all $t \geq 1$. Specifically, the *state sequence* (s_t), generated by the mechanism (M_t, G_t, H_t) is defined by

$$s_0 = s, \ s_t = \nu_t^1(e_t, s_{t-1}) \quad \text{for } t \geq 1 \ .$$

The *allocation sequence* (a_t) generated by the mechanism (M_t, G_t, H_t) is defined by

$$a_t = \nu_t^2(e_t, s_{t-1}) \quad \text{for } t \geq 1 \ .$$

5b. Decentralization

The evolutionary mechanism defined in the previous subsection need not be decentralized. This is because in verifying whether a message is an equilibrium message in some period t, we did not rule out the possibility of a single agent having the full information, e_t, about the environment. We do so now by formally introducing the notion of decentralization of information.

Assume that for each t, the set of environments at that date, E_t, is defined by two independent pieces of information, which are held by two separate agents. Specifically, assume that

$$E_t = U_t \times F_t \ ,$$

where U_t is to be interpreted as a set of utility functions (with typical element u_t), and F_t a set of production functions (with typical element f_t).

The evolutionary mechanism (M_t, G_t, H_t) is said to be *decentralized* if there exist sequences (A_t, B_t) such that
(i) A_t and B_t map $E_t \times S_{t-1} \times M_t$ into $\Re^n$,
(ii) $G_t(u_t, f_t, s_{t-1}, m_t) = A_t(u_t, s_{t-1}, m_t) + B_t(f_t, s_{t-1}, m_t)$.
This definition follows Hurwicz and Weinberger (1990) closely. Their paper also contains alternative but equivalent ways of defining the concept of decentralization. The definition basically conveys the idea that a consumer's "response" (A_t) to a message (m_t) can utilize only the information which consumers have, namely his utility function and the previous period's state (which is treated as "common knowledge"); a similar remark applies to a producer's "response" (B_t) to a message (m_t).

5c. Evaluation of the Performance of a Mechanism

The performance of a mechanism is evaluated by setting up a goal correspondence which, loosely speaking, specifies a set of allocations judged to be "socially desirable." Formally, a *goal correspondence* is a mapping Q from $E \times S_0$ to subsets of A.

For each specification of an environment $e \in E$, and an initial state $s \in S_0$, Q specifies a set of allocation sequences (a_t), the attainment of which should be the aim of a constructed mechanism.

The mechanism (M_t, G_t, H_t) is said to *realize the goal correspondence* Q if for each $(e, s) \in E \times S_0$, the sequence of allocations, (a_t), generated by the mechanism belongs to $Q(e, s)$.

6. Decentralized Evolutionary Realization of Optimality

In this final section, we show that a rolling plan can be obtained through a suitably designed decentralized evolutionary mechanism. This establishes, in particular, the following possibility result: there is a decentralized evolutionary mechanism which realizes optimality as defined in section 2a.

We begin by specifying the set of environments. Let $\xi > 0$ be a fixed real number, and $\bar{k}$ be another fixed real number satisfying $0 < \bar{k} < \xi$. Define

$$\begin{aligned} F &= \{f\colon \Re_+ \to \Re_+ \mid f \text{ satisfies (A.1)–(A.5) and} \\ &\qquad f(\bar{k}) \geq \bar{k},\ f(\xi) < \xi,\ f'(\bar{k}) < 1\}, \\ U &= \{u\colon \Re_+ \to \Re \mid u \text{ satisfies (A.6)–(A.9)}\}. \end{aligned}$$

We now specify $E_t = U \times F$ for all $t \geq 1$ and $E = \{(u, f)^\infty : (u, f) \in U \times F\}$. Thus, our intertemporal framework is interpreted as one with one consumer (with time-stationary utility function $u \in U$) and one producer (with time-stationary production function $f \in F$).

Notice that given any $f \in F$, there is a golden-rule stock, $\hat{k}$, and a maximum sustainable stock, k. (This follows, as in Section 2a, from assumptions (A.1)–(A.5).) Furthermore, our definition of F ensures that $\hat{k} < \bar{k} \leq k < \xi$.

We specify the space of allocations by $A_t = \Re_+$ for all $t \geq 1$, and the state space by $S_t = (0, \bar{k})$ for all $t \geq 0$. Thus, a typical allocation in period t should be interpreted as the consumption in that period; a typical state in period t will be the input stock at the end of that period.

Proposition 6.1: There is a decentralized evolutionary mechanism, such that if the initial state is $\mathbf{x} \in (0, \bar{k})$, then the state sequence (x_t^*) generated by the mechanism is the rolling plan from $\mathbf{x}$.

Proof: We specify the required mechanism, and simply check that it satisfies the property stated in the Proposition.

Define the message space in period t, $M_t = \Re^4_{++}$ for all $t \geq 1$; we write (suggestively) the typical element of M_t as

$$m_t = (x_t, c_t, d_{t+1}, r_t) .$$

The verification function in period t, G_t, is specified by defining A_t and B_t as follows for $t \geq 1$:

$$A_t(u, x_{t-1}, m_t) = \begin{bmatrix} [u'(c_t)/u'(d_{t+1})] - r_t \\ 0 \\ 0 \\ 0 \end{bmatrix},$$

$$B_t(f, x_{t-1}, m_t) = \begin{bmatrix} 0 \\ x_t + c_t - f(x_{t-1}) \\ x_{t-1} + d_{t+1} - f(x_t) \\ f'(x_t) - r_t \end{bmatrix}.$$

Now, $G_t(u, f, x_{t-1}, m_t)$ is defined as $A_t(u, x_{t-1}, m_t) + B_t(f, x_{t-1}, m_t)$ for $t \geq 1$. Note that, given (u, f) and $x_{t-1} \in (0, \bar{k})$, there is a unique maximal 2-plan from x_{t-1}, given by $(x_{t-1}, x_t^*, x_{t-1})$ with associated consumption sequence (c_t^*, d_{t+1}^*) satisfying $c_t^* = f(x_{t-1}) - x_t^*$ and $d_{t+1}^* = f(x_t^*) - x_{t-1}$. Furthermore, by the Ramsey-Euler equations, we have $[u'(c_t^*)/u'(d_{t+1}^*)] = f'(x_t^*)$. Thus, if we define $r_t^* = f'(x_t^*)$, and consider the message $m_t^* \equiv (x_t^*, c_t^*, d_{t+1}^*, r_t^*)$ we note that $A_t(u, x_{t-1}, m_t^*) = B_t(f, x_{t-1}, m_t^*) = 0$, so $G_t(u, f, x_{t-1}, m_t^*) = 0$; that is, m_t^* is an equilibrium message in period t.

Consider, next, any equilibrium message $\tilde{m} \equiv (\tilde{x}_t, \tilde{c}_t, \tilde{d}_{t+1}, \tilde{r}_t)$. Then $G_t(u, f, x_{t-1}, \tilde{m}_t) = 0$, and so $A_t(u, x_{t-1}, \tilde{m}_t) = 0 = B_t(f, x_{t-1}, \tilde{m}_t)$. This means that we have

$$\begin{aligned} u'(\tilde{c}_t)/u'(\tilde{d}_{t+1}) &= \tilde{r}_t , \\ \tilde{x}_t + \tilde{c}_t &= f(x_{t-1}) , \\ x_{t-1} + \tilde{d}_{t+1} &= f(\tilde{x}_t) , \\ f'(\tilde{x}_t) &= \tilde{r}_t . \end{aligned}$$

Thus $(x_{t-1}, \tilde{x}_t, x_{t-1})$ is a 2-plan from x_{t-1} with associated consumption sequence $(\tilde{c}_t, \tilde{d}_{t+1})$, which satisfies $\tilde{c}_t > 0$, $\tilde{d}_{t+1} > 0$ and

$$u'(\tilde{c}_t)/u'(\tilde{d}_{t+1}) = f'(\tilde{x}_t) .$$

Then by Proposition 2.1, $(x_{t-1}, \tilde{x}_t, x_{t-1})$ is a maximal 2-plan from x_{t-1}. Since $(x_{t-1}, x_t^*, x_{t-1})$ is the unique maximal 2-plan from x_{t-1}

we have $\tilde{x}_t = x_t^*$, $\tilde{c}_t = c_t^*$, $\tilde{d}_{t+1} = d_{t+1}^*$ and $\tilde{r}_t = r_t^*$. That is, m_t^* is the unique equilibrium message. We have now checked that G_t, as defined above, is a verification function.

Finally, define the outcome function, H_t, as the map

$$H_t(x_t, c_t, d_{t+1}, r_t) = (x_t, c_t) \ .$$

For the equilibrium message m_t^*, we have by Lemma 3.1, $x_t^* < \bar{k}$, so $x_t^* \in S_t$ as required; also $c_t^* > 0$, so $c_t^* \in A_t$ as required. We have now demonstrated that (M_t, G_t, H_t) as defined above is a decentralized, evolutionary mechanism.

Consider period 1, with $\mathbf{x} \in (0, \bar{k})$ the state in period 0. Our above demonstration has shown that if the unique maximal 2-plan from $\mathbf{x}$ is $(\mathbf{x}, x_1^*, \mathbf{x})$, then x_1^* is the equilibrium state (generated by the mechanism) in period 1. This means that $x_1^* = h(\mathbf{x})$, where h is defined in Section 2c. Repeating this step for $t = 2, 3, \ldots$ shows that the state sequence (x_t^*) generated by the mechanism (M_t, G_t, H_t) satisfies $x_0^* = \mathbf{x}$, and $x_{t+1}^* = h(x_t^*)$ for $t \geq 0$. That is, it is the rolling plan from $\mathbf{x}$.

Q.E.D.

Define the optimality goal correspondence, Q, as follows. For each $(u, f, \mathbf{x}) \in U \times F \times (0, \bar{k})$, let

$$Q(u, f, \mathbf{x}) = \{(c_t)\text{: there is an optimal plan } (x_t) \text{ from } \mathbf{x}, \text{ whose associated consumption sequence is } (c_t)\} \ .$$

That is, the goal is to attain a plan which is optimal as defined in Section 2.1. We can now state our "possibility result" as follows.

Theorem 6.1: There is a decentralized evolutionary mechanism which realizes the optimality goal correspondence.

Proof: Consider the decentralized evolutionary mechanism (M_t, G_t, H_t) constructed in the proof of Proposition 6.1. Then, given any $(u, f) \in U \times F$, and any $\mathbf{x} \in (0, \bar{k})$, the sequence of states (x_t^*) generated by the mechanism is the rolling plan from $\mathbf{x}$. Thus, the sequence of allocations (c_t^*) generated by the mechanism is the consumption sequence associated with the rolling plan from $\mathbf{x}$. By Theorem 4.3, the rolling plan from $\mathbf{x}$ is an optimal plan from $\mathbf{x}$. Hence, the sequence of allocations (c_t^*) generated by the mechanism belongs to $Q(u, f, \mathbf{x})$. That is, (M_t, G_t, H_t) realizes the optimality goal correspondence.

Q.E.D.

Remarks:

(i) We make the following somewhat informal observation about the verification functions, G_t, which appear in our constructed mechanism.

Notice that in period t, given the message $m_t = (x_t, c_t, d_{t+1}, r_t)$ the consumer is asked to verify

$$u'(c_t)/u'(d_{t+1}) = r_t \ .$$

The consumer, knowing u, can surely do this. The condition to be verified is simply the equality of the marginal rate of intertemporal substitution on the consumption side with an appropriate "shadow" price ratio, r_t.

The producer is asked to verify

$$\begin{aligned} x_t + c_t &= f(x_{t-1}) \ , \\ x_{t-1} + d_{t+1} &= f(x_t) \ , \\ f'(x_t) &= r_t \ . \end{aligned}$$

The producer, knowing f and the previous period's input stock x_{t-1}, can do this. The first two conditions are to be interpreted as verification of feasibility. The third condition is simply the equality of the marginal rate of transformation on the production side with an appropriate "shadow" price ratio, r_t.

The above verifications imply that the following conditions hold:

$$\begin{aligned} x_t + c_t &= f(x_{t-1}) \ , \\ x_{t-1} + d_{t+1} &= f(x_t) \ , \\ u'(c_t)/u'(d_{t+1}) &= f'(x_t) \ . \end{aligned}$$

But these conditions constitute a complete characterization of the maximal 2-plan from x_{t-1}, according to Proposition 2.1. Thus, the equilibrium state of the mechanism in period t is precisely the same as $h(x_{t-1})$, the rolling plan input in period t.

(ii) If (m_t) is the sequence of equilibrium messages of our constructed mechanism (given u, f and $\mathbf{x}$), then

$$r_t = f'(x_t) \ ,$$

where (x_t) is the sequence of equilibrium states generated by the mechanism, and therefore also the rolling plan from $\mathbf{x}$. If we define

$$p_0 = 1 \text{ and } p_{t+1} = (p_t/r_t) \quad \text{for } t \geq 0 \ ,$$

then we have for $t \geq 0$

$$p_{t+1}f(x_t) - p_t x_t \geq p_{t+1}f(x) - p_t x \quad \text{for all } x \geq 0 .$$

That is, (p_t) is a sequence of intertemporal profit maximizing prices, the shadow prices that Malinvaud (1953) was concerned with.

The shadow prices that Gale and Sutherland (1968) are concerned with would, in addition, have to satisfy for $t \geq 1$

$$u(c_t) - p_t c_t \geq u(c) - p_t c \quad \text{for all } c \geq 0 .$$

This in our framework would require

$$[u'(c_t)/u'(c_{t+1})] = f'(x_t) \quad \text{for } t \geq 1 ,$$

a condition which is *not* satisfied by the rolling plan (x_t) unless $\mathbf{x} = \hat{k}$, the golden-rule input stock. That is, while for the maximal 2-plan formulated in period t, we have

$$[u'(c_t)/u'(d_{t+1})] = f'(x_t) ,$$

the "second-period consumption" of the maximal 2-plan formulated in period t, d_{t+1}, is not carried out in period $(t+1)$; instead the "first-period consumption" of the maximal 2-plan formulated in period $(t+1)$, c_{t+1}, is carried out in period $(t+1)$.

(iii) An "undiscounted" optimality notion which has figured prominently in the literature is the "overtaking" criterion of Atsumi (1965) and von Weizsäcker (1965); this was subsequently refined by Gale (1967) in terms of the "catching up" criterion. If a plan is "catching-up optimal," it is both efficient and good, and hence maximizes long-run average utility of consumption. That is, it is optimal in the sense of our definition in Section 2a. The converse is not true; the rolling plan from any $\mathbf{x} \in (0, k)$ with $\mathbf{x} \neq \hat{k}$ is optimal in our sense, but is not "catching-up optimal."

We conjecture, but do not attempt to prove, that an analogue of the Hurwicz-Weinberger result can be shown for the undiscounted case; that is, it is impossible for a decentralized evolutionary mechanism to realize the "catching-up optimality" criterion.

References

Arrow, K. J., and Hahn, F. (1971): *General Competitive Analysis*. San Francisco: Holden-Day.

Atsumi, H. (1965): "Neoclassical Growth and Efficient Program of Capital Accumulation." *Review of Economic Studies* 32: 127–136.

Bhattacharya, R. N., and Majumdar, M. (1989): "Controlled Semi-Markov Models under Long-Run Average Rewards." *Journal of Statistical Planning and Inference* 22: 223–242.

Birkhoff, G., and MacLane, S. (1977): *A Survey of Modern Algebra*. New York: Macmillan.

Blackwell, D. (1962): "Discrete Dynamic Programming." *Annals of Mathematical Statistics* 33: 719–726.

Brock, W. A. (1971): "Sensitivity of Optimal Growth Paths with Respect to a Change in Target Stocks." In *Contributions to the Von Neumann Growth Model*, edited by G. Bruckmann and W. Weber. New York: Springer.

Brock, W. A., and Majumdar, M. (1988): "On Characterizing Optimal Competitive Programs in Terms of Decentralizable Conditions." *Journal of Economic Theory* 45: 262–273.

Cass, D. (1972): "On Capital Overaccumulation in the Aggregative Neo-Classical Model of Economic Growth: A Complete Characterization." *Journal of Economic Theory* 4: 200–223.

Dasgupta, A. (1964): "A Note on Optimum Savings." *Econometrica* 32: 431–433.

Dasgupta, S., and Mitra, T. (1988a): "Characterization of Intertemporal Optimality in Terms of Decentralizable Conditions: The Discounted Case." *Journal of Economic Theory* 45: 247–287.

—— (1988b): "Intertemporal Optimality in a Closed Linear Model of Production." *Journal of Economic Theory* 45: 288–315.

Dutta, P. K. (1986): "Essays in Intertemporal Allocation Theory." Ph. D. Thesis, Cornell University, Ithaca, New York.

—— (1989): "What Do Discounted Optima Converge to? A Theory of Discount Rate Asymptotics in Economic Models." Discussion Paper No. 426, Columbia University.

Flynn, J. (1976): "Conditions for the Equivalence of Optimality Criteria in Dynamic Programming." *Annals of Mathematical Statistics* 4: 936–953.

Gale, D. (1967): "On Optimal Development in a Multi-Sector Economy." *Review of Economic Studies* 34: 1–18.

Gale, D., and Sutherland, W. R. S. (1968): "Analysis of a One-Good Model of Economic Development." In *Mathematics of the Decision Sciences, Part 2*, edited by G. B. Dantzig and A. F. Veinott, Jr. (Lectures in Applied Mathematics 12). Providence: American Mathematical Society.

Goldman, S. M. (1968): "Optimal Growth and Continual Planning Revision." *Review of Economic Studies* 35: 145–154.

Hahn, F. (1989): "Auctioneer." In *The New Palgrave: General Equilbrium*, edited by J. Eatwell, M. Milgate and P. Newman. London: Macmillan.

Howard, R. (1960): *Dynamic Programming and Markov Processes*. Cambridge, Mass.: MIT Press.

Hurwicz, L. (1986): "On Intertemporal Decentralization and Efficiency in Resource Allocation Mechanisms." *Studies in Mathematics* 25: 238–350.

Hurwicz, L., and Majumdar, M. (1988): "Optimal Intertemporal Allocation Mechanisms and Decentralization of Decisions." *Journal of Economic Theory* 45: 228–261.

Hurwicz, L., and Weinberger, H. (1990): "A Necessary Condition for Decentralization and an Application to Intertemporal Allocation." *Journal of Economic Theory* 51: 313–345.

Jeanjean, P. (1974): "Optimal Development Programs under Uncertainty: The Undiscounted Case." *Journal of Economic Theory* 7: 66–92.

Koopmans, T. C. (1957): *Three Essays on the State of Economic Science*. New York: McGraw Hill.

Majumdar, M. (1988): "Decentralization in Infinite Horizon Economies: An Introduction." *Journal of Economic Theory* 45: 217–227.

Malinvaud, E. (1953): "Capital Accumulation and Efficient Allocation of Resources." *Econometrica* 21: 233–268.

Mount, K., and Reiter, S. (1974): "The Informational Size of Message Spaces." *Journal of Economic Theory* 8: 161–192.

Pigou, A. C. (1928): *The Economics of Welfare*. London: Macmillan.

Ramsey, F. (1928): "A Mathematical Theory of Savings." *Economic Journal* 38: 543–559.

Rawls, J. (1971): *A Theory of Justice*. Cambridge, Mass.: Belknap Harvard.

Ross, S. (1968): "Arbitrary State Markovian Decision Processes." *Annals of Mathematical Statistics* 39: 2118–2122.

Samuelson, P. A. (1958): "An Exact Consumption-Loan Model of Interest With or Without the Social Contrivance of Money." *Journal of Political Economy* 66: 467–482.

Scarf, H. (1960): "Some Examples of Global Instability of the Competitive Equilibrium." *International Economic Review* 1: 157–172.

Smale, S. (1989): "Global Analysis in Economic Theory." In *The New Palgrave: General Equilibrium*, edited by J. Eatwell, M. Milgate and P. Newman. London: Macmillan.

Veinott Jr., A. F. (1966): "On Finding Optimal Policies in Discrete Dynamic Programming with no Discounting." *Annals of Mathematical Statistics* 37: 1284–1294.

Weizsäcker, C. C. von (1965): "Existence of Optimal Programs of Accumulation for an Infinite Time Horizon." *Review of Economic Studies* 32: 85–104.

10
Complements and Details

MUKUL MAJUMDAR

Department of Economics, Cornell University

Here I sketch some of the proofs left out in the published version, add a few comments and report on some new results. First, let me start with some details.

I. "DECENTRALIZATION IN INFINITE HORIZON ECONOMIES: AN INTRODUCTION"

The proofs of Propositions 2 and 3 will now be spelled out. In referring to an equation in the original article, I use the same number.

Proof of Proposition 2

Since the competitive program is interior, one actually has

$$\bar{p}_t = \delta^t u'(\bar{c}_t), \qquad \bar{p}_{t+1} f'(\bar{x}_t) = \bar{p}_t \tag{N.2.1}$$

and the 'Ramsey-Euler' condition:

$$\delta u'(\bar{c}_{t+1}) f'(\bar{x}_t) = (u'\bar{c}_t) \tag{R-E}$$

Three cases have to be considered:

Case I: $\bar{y}_0 = y_\delta^*$;
Case II: $\bar{y}_0 < y_\delta^*$;
Case III: $\bar{y}_0 > y_\delta^*$;

Case I: $\bar{y}_0 = y_\delta^*$. One has to show that $\bar{x}_t = x_\delta^*$, $\bar{y}_t = y_\delta^*$ for all t. Suppose that '$\bar{c}_0 < c_\delta^*$'; then one has '$\bar{x}_0 > x_\delta^*$', implying $\bar{y}_1 = f(\bar{x}_0) > f(x_\delta^*) = y_\delta^*$. By condition (*S*), $\bar{c}_1 \geq c_\delta^*$. But $f'(\bar{x}_0) < f'(x_\delta^*)$ and, by (*R-E*), $u'(\bar{c}_1) > \delta u'(\bar{c}_1) f'(\bar{x}_0) = u'(\bar{c}_0) > u'(c_\delta^*)$.Hence, $\bar{c}_1 < c_\delta^*$, a contradiction. Similarly, $\bar{c}_0 > c_\delta^*$ leads to a contradiction. Hence, $\bar{c}_0 = c_\delta^*$, $\bar{x}_0 = x_\delta^*$; and, $\bar{y}_1 = y_\delta^*$. Complete the proof by an induction argument.

Case II: $\bar{y}_0 < y_\delta^*$. One shows the following:

$$\bar{y}_t < y_\delta^* \text{ and } \bar{c}_{t+1} > \bar{c}_t \text{ for all } t. \qquad (N.2.2)$$

Assuming $(N.2.2)$, we can complete the proof by noting simply that $u'(\bar{c}_t) < u'(\bar{c}_0)$ for all t implies that (a) $\bar{p}_t$ goes to zero when $\delta < 1$. (b) $\bar{p}_t$ is bounded when $\delta = 1$. We shall now sketch a proof of $(N.2.2)$.

Step I: By condition (S), $\bar{p}_0 \geq p_0^*$, i.e., $\bar{c}_0 \leq c_\delta^*$.

Step II: $\bar{x}_0 \leq x_\delta^*$.

Proof If $\bar{x}_0 > x_\delta^*, f'(\bar{x}_0) < f'(x_\delta^*)$ and $\bar{y}_1 > y_\delta^*$. By (S), $u'(\bar{c}_1) \leq u'(c_\delta^*)$. By $(R\text{-}E)$, $u'(\bar{c}_1) = u'(\bar{c}_0)/\delta f'(\bar{x}_0) > u'(c_\delta^*)$, a contradiction.

Step III: Show that the following cannot occur:

$$(i) \quad \bar{c}_0 = c_\delta^*, \qquad \bar{x}_0 = x_\delta^*.$$

$$(ii) \quad \bar{c}_0 < c_\delta^*, \qquad \bar{x}_0 = x_\delta^*.$$

$$(iii) \quad \bar{c}_0 = c_\delta^*, \qquad \bar{x}_0 < x_\delta^*.$$

Proof (i): contradicts $\bar{y}_0 < y_\delta^*$; consider (ii): here, $\bar{y}_1 = y_\delta^*$. By $(R\text{-}E)$, $u'(\bar{c}_1) = u'(\bar{c}_0)/\delta f'(\bar{x}_0) = u'(\bar{c}_0)/\delta f'(x_\delta^*) = u'(\bar{c}_0) > u'(c_\delta^*)$. Hence $\bar{c}_1 < c_\delta^*$, and $\bar{x}_1 > x_\delta^*$. This implies $\bar{y}_2 > y_\delta^*$; by (S), $u'(\bar{c}_2) \leq u'(c_\delta^*)$. But, by $(R\text{-}E)$, $u'(\bar{c}_2) = u'(\bar{c}_1)/\delta f'(\bar{x}_1) > u'(c_\delta^*)$, a contradiction.

Finally, (iii): here, $\bar{y}_1 < y_\delta^*$. By (S) $u'(\bar{c}_1) \geq u'(c_\delta^*)$. But by $(R\text{-}E)$, $u'(\bar{c}_1) = u'(\bar{c}_0)/\delta f'(\bar{x}_0) < u'(c_\delta^*)$, a contradiction.

Step IV: From the earlier steps, $\bar{c}_0 < c_\delta^*$ and $\bar{x}_0 < x_\delta^*$. Hence $\bar{y}_1 < y_\delta^*$. Also, $u'(\bar{c}_1) = u'(\bar{c}_0)/\delta f'(\bar{x}_0) < u'(\bar{c}_0)$; or $\bar{c}_1 > \bar{c}_0$. This verifies $(N.2.2)$ for $t = 1$. An induction argument completes the proof of $(N.2.2)$.

Case III. $\bar{y}_0 > y_\delta^*$. An argument parallel to Case II disposes of this case.

(Q.E.D.)

Proof of Proposition 3

If $(\mathbf{x},\mathbf{y},\mathbf{c})$ is optimal, it is competitive and is monotonic in the sense of (2.6) (see, e.g., Majumdar and Nermuth 1982). Suppose that $(\mathbf{x},\mathbf{y},\mathbf{c})$ is competitive and satisfies (2.6). First, we show

Step I:

$$x_0 < x_\delta^* \text{ implies } x_t < x_\delta^* \text{ for all } t \geq 0 \qquad (N.2.3)$$

$$x_0 > x_\delta^* \text{ implies } x_t > x_\delta^* \text{ for all } t \geq 0 \qquad (N.2.4)$$

Proof Suppose $T+1$ is the first period such that $x_0,\ldots,x_T < x_\delta^*$ and $x_{T+1} \geq x_\delta^*$. Then $\delta f'(x_{T+1}) \leq 1$. Since $(\mathbf{x},\mathbf{y},\mathbf{c})$ is competitive, one has $\delta u'(c_{T+2})f'(x_{T+1}) = u'(c_{T+1})$, implying $c_{T+2} \leq c_{T+1}$. But, $x_{T+1} > x_T$ implies that $f(x_{T+1}) > f(x_T)$ or

$c_{T+2} + x_{T+2} > c_{T+1} + x_{T+1}$ and, by (2.6), $x_{T+2} \le x_{T+1}$; hence, $c_{T+2} > c_{T+1}$, a contradiction establishing (*N*.2.3). Similarly (*N*.2.4) can be established.

Step II:

$$x_0 = x^*_\delta \text{ implies } x_t = x^*_\delta \text{ for all } t \ge 0 \qquad (N.2.5)$$

In order to complete the proof of **Proposition 3**, assume, first, that $x_0 \le x^*_\delta$. Then $x_t \le x^*_\delta$ for all t, and $x_{t+1} \ge x_t$ for all t. Let $\hat{x}$ be the limit of (x_t). Clearly $\hat{x} \le x^*_\delta$. Also, $\hat{x} \ge x_0 > 0$. We know that (c_t) converges to $\hat{c} = f(\hat{x}) - \hat{x}$, and $\hat{c} > 0$. When (i) $\delta < 1$, the sequence $(\delta^t u'(c_t)x_t)$ goes to zero as t goes to infinity; (ii) $\delta = 1$, $p_t x_t$ is bounded, implying optimality of (**x**,**y**,**c**). The proof is completed by applying similar arguments in the case $x_0 > x^*_\delta$.

(Q.E.D.)

II. Impossibility of Decentralization: The Undiscounted Case

As remarked in the **Introduction**, an impossibility theorem analogous to (T.5.2) of Hurwicz-Majumdar (to be referred to as **H-M**) continues to hold even when one considers the "undiscounted" case where the *Ramsey-Weizsäcker* overtaking criterion is used to evaluate alternative feasible programs. Here, a precise statement and a sketch of the proof of such an impossibility theorem are presented (see Majumdar 1991 for more details). The model developed in **H-M** is used, and all the assumptions relevant for (T.5.2) are maintained.

A feasible $(\mathbf{x}^*,\mathbf{y}^*,\mathbf{c}^*)_y$ is *Ramsey-Weizsäcker* optimal if

$$\limsup_{T\to\infty} \left[\sum_{t=0}^{T} u(c_t) - u(c_t^*)\right] \le 0 \qquad (N.3.1)$$

for all feasible $(\mathbf{x},\mathbf{y},\mathbf{c})_y$.

Let E be the set of all *admissible environments*, i.e.,

$$\mathrm{E} = \{e \equiv (\mathbf{f},u,y) : \text{there is a unique optimal program given } e\}.$$

Now define the *Ramsey-Weizsäcker evaluation criterion* O^* from E into $S^+ \times S^+ \times S^+$ as

$$O^*(e) = (\mathbf{x}^*,\mathbf{y}^*,\mathbf{c}^*)_y \qquad (N.3.2)$$

One can prove the following:

Impossibility Theorem. *The Ramsey-Weizsäcker criterion O^* is sensitive to a change in the technology in every period $T \ge 2$; hence, it is not decentralizable.*

Proof of the Impossibility Theorem

The main steps of the proof will be sketched. Observe that the proofs of R.4.1 and T.4 of **H-M** can be used in the present context. Thus, one has to establish only the sensitivity property of O^*. Consider, first, any stationary environment $e = (f^{(\infty)}, u, y)$ where f satisfies (F.1)–(F.4). Let x_g be the unique positive solution to the equation.

$$f'(x) = 1.$$

Write $y_g \equiv f(x_g)$, $c_g \equiv y_g - x_g$. The triplet (x_g, y_g, c_g) represents the golden rule input, stock and consumption respectively.

Recall, from Gale (1967), that to simplify notation one can set $u(c_g) = 0$ and say that a feasible $(\mathbf{x},\mathbf{y},\mathbf{c})_y$ is *good* if there exists a *finite* M ($> -\infty$) such that

$$\sum_{t=0}^{T} u(c_t) \geq M \text{ for all } T \geq 0 \qquad (N.3.3)$$

A feasible $(\mathbf{x},\mathbf{y},\mathbf{c})_y$ is *bad* if

$$\lim_{T \uparrow \infty} \sum_{t=0}^{T} u(c_t) = -\infty \qquad (N.3.4)$$

One can show (see Majumdar-Mitra 1982, Section 4) the following:

L.1 *There exists a good program.*

L.2 *If a program is not good, it is bad.*

L.3 *If* $(\mathbf{x},\mathbf{y},\mathbf{c})_y$ *is good,*

$$\sum_{t=0}^{\infty} u(c_t) \qquad (N.3.5)$$

is well defined.

L.4 *There exists a unique optimal program* $(\mathbf{x}^*,\mathbf{y}^*,\mathbf{c}^*)_y$.

It follows from L.2 and L.4 that the optimal $(\mathbf{x}^*,\mathbf{y}^*,\mathbf{c}^*)_y$ is necessarily good; define for each $y > 0$

$$V(y) = \sum_{t=0}^{\infty} u(c_t^*) \qquad (N.3.6)$$

Now verify that

L.5 a. *V is increasing on P, i.e., "$y_1 > y_2$" implies "$V(y_1) > V(y_2)$".*

b. *V is strictly concave on P.*

It follows that (see, e.g., Nikaido 1968, Theorem 3.14).

L.6 *V is continuous on P.*

The following "optimality principle" can also be verified directly:

L.7 *For any* $T \geq 1$, $< (x_t^*), (y_t^*), (c_t^*) >_{t=T}^{\infty}$ *is the unique optimal program from* y_T^*.

L.7 leads to the following: for any $t \geq 0$,

$$V(y_t^*) = u(c_t^*) + V(y_{t+1}^*) \tag{N.3.7}$$

In particular, for $t = 0$

$$V(y) = u(c_0^*) + V[f(y - c_0^*)] \tag{N.3.8}$$

Consider the following two problems: for a given $y > 0$

$$\text{maximize "}u(c) + V[f(y - c)]\text{"}$$

$$\text{subject to "}0 \leq c \leq y\text{"} \tag{N.3.9}$$

$$\text{or, subject to "}0 < c < y.\text{"} \tag{N.3.9'}$$

A careful argument is needed to guarantee that the maximum is attained at a unique point $\bar{c}$ satisfying $0 < \bar{c} < y$ in both problems. One can then, depending on the definition of u, derive the dynamic programming equation satisfied by V:

$$V(y) = \max_{0 \leq c \leq y} \{u(c) + V[f(y - c)]\} \tag{N.3.10}$$

$$\text{or,} \quad V(y) = \max_{0 < c < y} \{u(c) + V[f(y - c)]\} \tag{N.3.11}$$

One can next verify that the maximum on the right side of (N.3.10) or (N.3.11) is, in fact, attained at c_0^*, i.e., $\bar{c} = c_0^*$.

Define the function $h : P \to P$ as:

$$h(y) \equiv c_0^* \tag{N.3.12}$$

and define $i : P \to P$ as:

$$i(y) \equiv y - h(y).$$

The functions h and i are *the optimal consumption policy function* and *the optimal investment policy function* respectively.

L.8 (Mirman and Zilcha 1977, Lemma 2) *V is differentiable at* $y > 0$, $V'(y) = u'(h(y))$.

The Ramsey-Euler condition characterizing optimality can be stated as:

$$u'(h(y)) = f'(i(y))u'[h(f(i(y)))] \tag{N.3.13}$$

The proof of the undiscounted case can be completed by adapting the steps of the discounted case. The relevant properties of h and i needed to exploit the ideas in the proof of the discounted case can be verified easily. **(Q.E.D.)**

III. An Alternative Period-by-Period Characterization of Optimality

The verification rule (S) identifying the optimality of competitive programs involves the parameters (y_δ^*, p_t^*). The structure of the aggregative model can be exploited to specify a verification rule that dispenses with these parameters. Given a competitive program $(\mathbf{x},\mathbf{y},\mathbf{c};\mathbf{p})$ from $y_0 > 0$, define the *current price sequence* $q = (q_t)$ by

$$q_t = p_t/\delta^t \quad \text{for all} \quad t \geq 0 \tag{N.4.1}$$

To be sure, in the undiscounted case $\delta = 1$, $q_t = p_t$ for all $t \geq 0$. Now consider the following period-by-period verification rule:

$$(q_{t+1} - q_t)(y_{t+1} - y_t) \leq 0 \quad \text{for all} \quad t \geq 0 \tag{S'}$$

In contrast with the earlier condition (S), this condition (S') does not involve the explicit appearance of the golden rule (or modified golden rule) stock y_δ^* or the golden rule support prices p_t^*. It will be shown that a competitive program satisfies the condition (S') if and only if it is optimal.

It is worth noting that a competitive program $(\mathbf{x},\mathbf{y},\mathbf{c};\mathbf{p})$ from y_0 satisfies (S') if and only if it satisfies

$$(q_{t+1} - q_t)(x_{t+1} - x_t) \leq 0 \quad \text{for all} \quad t \geq 0 \tag{S''}$$

To verify this equivalence, using the condition (G) we get for $t \geq 0$,

$$u(c_t) - q_t c_t \geq u(c_{t+1}) - q_t c_{t+1}$$

$$u(c_{t+1}) - q_{t+1} c_{t+1} \geq u(c_t) - q_{t+1} c_t$$

Adding the inequalities and transposing terms, we get

$$(q_{t+1} - q_t)(c_{t+1} - c_t) \leq 0 \quad \text{for all} \quad t \geq 0 \tag{N.4.2}$$

Thus, if (S'') is satisfied, adding the inequalities in (S'') and (N.4.2), one obtains (S'). To go in the other direction, observe that by condition (M), we get for all $t \geq 0$,

$$\delta q_{t+1} f(x_t) - q_t x_t \geq \delta q_{t+1} f(x_{t+1}) - q_t x_{t+1} \quad (N.4.3)$$

$$\delta q_{t+2} f(x_{t+1}) - q_{t+1} x_{t+1} \geq \delta q_{t+2} f(x_t) - q_{t+1} x_t \quad (N.4.4)$$

Adding the inequalities (N.4.3) and (N.4.4), we get—through a transposition of terms—the following:

$$\delta(q_{t+2} - q_{t+1})(y_{t+2} - y_{t+1}) \geq (q_{t+1} - q_t)(x_{t+1} - x_t) \quad \text{for all} \quad t \geq 0 \quad (N.4.5)$$

Thus, if (S') holds, it follows from (N.4.5) that (S'') is satisfied as well.

Proposition (N.1)

A *competitive program* (**x**,**y**,**c**;**p**) *from* $y_0 > 0$ *which satisfies* (S'), *is optimal.*

Proof The proof relies on two basic steps. One first verifies

Step I:

(*i*) *If, for some* $t \geq 0$, $x_t < x_\delta^*$, *then* $x_{t+1} \geq x_t$ *and* $y_{t+1} \geq y_t$.

(*ii*) *If, for some* $t \geq 0$, $x_t > x_\delta^*$, *then* $x_{t+1} \leq x_t$ *and* $y_{t+1} \leq y_t$.

Proof Consider case (*i*). Suppose that for a competitive program satisfying (S'), $x_t < x_\delta^*$ for some period $t \geq 0$. Then $\delta f'(x_t) > 1$. Using (R-E), $q_{t+1} = [q_t/\delta f'(x_t)] < q_t$. Using this in ($S'$), one obtains $y_{t+1} \geq y_t$. From (S'') one gets $x_{t+1} \geq x_t$.

The proof of case (*ii*) is similar.

Step II: (x_t, y_t, c_t) *converges to* $(x_\delta^*, y_\delta^*, c_\delta^*)$ *as t tends to infinity.*

Suppose $y_0 < y_\delta^*$. First we verify that

$$x_t < x_\delta^* \quad \text{for all} \quad t \geq 0. \quad (N.4.6)$$

To verify (N.4.6), first we establish the following claims:

L.9 *Suppose* (**x**,**y**,**c**;**p**) *is a competitive program from* y_0.

(*i*) *If for some* $s \geq 0$, *we have* $y_{s+1} > y_s$ *and* $x_s \geq x_\delta^*$, *then* $x_{t+1} > x_t$ *and* $y_{t+1} > y_t$ *for all* $t \geq s$.

(*ii*) *If for some* $s \geq 0$, *we have* $y_{s+1} < y_s$ *and* $x_s \leq x_\delta^*$, *then* $x_{t+1} < x_t$ *and* $y_{t+1} < y_t$ *for all* $t \geq s$.

Proof of L.9 Consider claim (i).

Since $x_s \geq x_\delta^*$, $\delta f'(x_s) \leq 1$. Using (R-E) we get

$$u'(c_s) = \delta f'(x_s)u'(c_{s+1}) \leq u'(c_{s+1})$$

so that $c_s \geq c_{s+1}$. Hence,

$$y_s - x_s \geq y_{s+1} - x_{s+1} > y_s - x_{s+1}.$$

This leads to $x_{s+1} > x_s \geq x_\delta^*$. So

$$y_{s+2} = f(x_{s+1}) > f(x_s) = y_{s+1}.$$

Thus, $y_{s+2} > y_{s+1}$ and $x_{s+1} > x_\delta^*$. This argument can be repeated to establish (i). The proof of (ii) is similar.

If (N.4.6) is violated, let s be the first period for which $x_s \geq x_\delta^*$. If $s = 0$, we have $y_1 = f(x_0) \geq f(x_\delta^*) = y_\delta^* > y_0$. If $s \geq 1$, $x_{s-1} < x_\delta^*$ so that $y_{s+1} = f(x_s) \geq f(x_\delta^*) > f(x_{s-1}) = y_s$. In either case, $y_{s+1} > y_s$. By using (L.9), $x_{s+1} > x_s \geq x_\delta^*$ and $y_{s+2} > y_{s+1}$. This contradicts **Step I**. Hence, (N.4.6) is verified.

Now, from (N.4.6) and **Step I**, we conclude that $x_{t+1} \geq x_t$ and $x_t < x_\delta^*$ for all $t \geq 0$. Hence, as t tends to infinity, the sequence (x_t) converges to some $\hat{x}$ satisfying $0 < \hat{x} \leq x_\delta^*$.

Also, the sequence $c_t \equiv [f(x_{t-1}) - x_t]$ converges to $\hat{c} \equiv [f(\hat{x}) - \hat{x}] > 0$. By using the condition (R-E), we get $\delta f'(\hat{x}) = 1$ so that $\hat{x} = x_\delta^*$. Hence, c_t converges to c_δ^* and y_t converges to y_δ^*. When $y_0 \geq y_\delta^*$, use a similar argument to complete the proof of **Step II**.

When $\delta = 1$, from **Step II**, we get that $p_t = u'(c_t)$ converges to $u'(c_\delta^*) \equiv p_\delta^*$ as t tends to infinity. Also the convergent sequence (x_t) is bounded. Hence $p_t x_t$ is bounded [recall (2.3a)]. When $0 < \delta < 1$, **Step II** implies that $q_t = u'(c_t)$ converges to $u'(c_\delta^*)$ at t tends to infinity. Hence, the sequence $p_t \equiv \delta^t q_t$ converges to zero as t tends to infinity. Since (x_t) is bounded, $\lim_{t \uparrow \infty} p_t x_t = 0$ [recall (2.3.b)].

In both cases, optimality of (**x**,**y**,**c**;**p**) follows from **Proposition 1** in the **Introduction**. This completes the proof of **Proposition (N.1)**. **(Q.E.D.)**

We now state a converse of **Proposition (N.1)**.

Proposition (N.2)

Let (**x**,**y**,**c**;**p**) *be a competitive program from* $y_0 > 0$ *such that* (**x**,**y**,**c**;**p**) *is optimal from* y_0. *Then* (**x**,**y**,**c**;**p**) *satisfies* (S').

Proof of Proposition (N.2)

Since (**x**,**y**,**c**;**p**) is competitive, $x_t > 0$, $y_t > 0$, $c_t > 0$ for all t. Hence, condition (G) implies that $u'(c_t) = q_t$ for all t.

Next, we recall the definition of the *value function*. For any $y > 0$, let $(\mathbf{x}^*,\mathbf{y}^*,\mathbf{c}^*)_y$ be the unique optimal program from y. In the *discounted case* ($0 < \delta < 1$), define

$$V(y) = \sum_{t=0}^{\infty} \delta^t u(c_t^*). \qquad (N.4.7)$$

In the undiscounted case [writing $u(c_g) = 0$] define [as in (N.3.6)]:

$$V(y) = \sum_{t=0}^{\infty} u(c_t^*). \qquad (N.4.8)$$

In either case, as in L.8,

L.10 *V is concave on P; V is differentiable at $y > 0$*

$$V'(y) = u'(c_0^*). \qquad (N.4.9)$$

Coming back to the given competitive program $(\mathbf{x},\mathbf{y},\mathbf{c};\mathbf{p})$ which is also optimal from y_0, it follows that for any $y > 0$ and any $t \in N$,

$$V(y) - V(y_t) \leq V'(y_t)(y - y_t) = q_t(y - y_t).$$

Transposing terms, we get for any $y > 0$ and any $t \in N$,

$$V(y_t) - q_t y_t \geq V(y) - q_t y. \qquad (N.4.10)$$

Pick any $s > 0$. Then using (N.4.10) for $t = s$, and $y = y_{s+1}$,

$$V(y_s) - q_s y_s \geq V(y_{s+1}) - q_s y_{s+1}. \qquad (N.4.11)$$

Using (N.4.10) for $t = s+1$ and $y - y_s$ we have

$$V(y_{s+1}) - q_{s+1} y_{s+1} \geq V(y_s) - q_{s+1} y_s \qquad (N.4.12)$$

Adding (N.4.11) and (N.4.12) and transposing terms again,

$$(q_{s+1} - q_s)(y_{s+1} - y_s) \leq 0.$$

Since the period $s \geq 0$ was arbitrarily picked, the condition (S') is established. **(Q.E.D.)**

It is not known whether the propositions (*N*.1) and (*N*.2) can be extended to multisector models; or to models where the technology represented by f is *productive*, i.e., $f(x)/x \geq \rho > 1$ for all $x > 0$.

This section is based on Majumdar and Mitra (1991). For a discussion of the problems related to decentralization in the presence of increasing returns, see Majumdar and Ray (1990).

REFERENCES

GALE, D. (1967): "On Optimal Development in a Multisector Economy", *Review of Economic Studies*, vol. 34, pp. 1–19.

MAJUMDAR, M., and T. MITRA (1982): "Intertemporal Allocation with a Non-Convex Technology: The Aggregative Framework", *Journal of Economic Theory*, vol. 27, pp. 101–136.

MAJUMDAR, M., and T. MITRA (1990): "Descentralizacion Intertemporal", *Cuadernos Economicos*, 46, 3, pp. 61–101. (The English translation of this paper is available as "Intertemporal Decentralization": Working Paper, 91-08, Center for Analytic Economics, 402 Uris Hall, Cornell University, Ithaca, NY 14853-7601.)

MAJUMDAR, M., and D. RAY (1990): "On Decentralization and Increasing Returns: Robinson Crusoe Revisited", in *Economic Theory and Policy*, B. Dutta, S. Gangopadhyay, D. Mookerjee and D. Ray, eds. Oxford University Press, pp. 69–87.

MAJUMDAR, M. (1991): "On Attaining *Ramsey-Weizsäcker* Optimality in a Decentralized Economy: An Impossibility Theorem" in *Essays in Economic Analysis and Policy*, D. Banerjee, ed. Oxford University Press, pp. 50–78.

MIRMAN, L., and I. ZILCHA (1977): "Characterizing Optimal Policies in a One-Sector Model of Economic Growth Under Uncertainty", *Journal of Economic Theory*, vol. 14, pp. 389–401.

NIKAIDO, H. (1968): *Convex Structures and Economic Theory*, Academic Press, New York.

About the Book and Editor

Our understanding of the efficient workings of decentralized price systems is severely tested by the problems presented when these systems are modelled without a definite terminal date. *Decentralization in Infinite Horizon Economies* brings together a set of the most important recent work on this topic along with new commentary and an excellent summary and overview by Mukul Majumdar. This book constitutes a definitive account of cutting-edge research on a topic of continuing importance in price theory.

Mukul Majumdar is H. T. Warshow and Robert Irving Warshow Professor of Economics at Cornell University and is a major contributor to the research on intertemporal resource allocation theory.

Contributors

Venkatesh Bala, McGill University
William A. Brock, University of Wisconsin
Swapan Dasgupta, Halifax University
Leonid Hurwicz, University of Minnesota
Mukul Majumdar, Cornell University
Tapan Mitra, Cornell University
Yaw Nyarko, New York University
Debraj Ray, Indian Statistical Institute
Hans Weinberger, University of Minnesota

Printed and bound by CPI Group (UK) Ltd, Croydon, CR0 4YY
23/10/2024
01778223-0012